Game Theory
and National Security

Game Theory and National Security

Steven J. Brams
and
D. Marc Kilgour

Basil Blackwell

Copyright © Steven J. Brams and D. Marc Kilgour 1988

First published 1988

Basil Blackwell Inc.
432 Park Avenue South, Suite 1503
New York, NY 10016, USA

Basil Blackwell Ltd
108 Cowley Road, Oxford, OX4 1JF, UK

British Library Cataloguing in Publication Data

Brams, Steven J. (Steven John)
Game theory and national security.
1. National security. Game theory
I. Title   II. Kilgour, D. Marc
327.1'1

ISBN 1–557–86003–3
ISBN 1–557–86004–1 Pbk

Library of Congress Cataloging in Publication Data

Game theory and national security /
Steven J. Brams and D. Marc Kilgour.
p.   cm.
Bibliography: p.
Includes index.
ISBN 1–557–86003–3 : $90.00
ISBN 1–557–86004–1 (pbk.) : $40.00
1. National security—Mathematical models.
2. War games. 3. Game theory.
I. Kilgour, D. Marc. II. Title
UA10.5.B73 1988
355'.03'0151—dc19

Typeset in 10 on 11½ pt Times
by Columns of Reading
Printed in the United States of America

Contents

Preface

Most applications of game theory to international relations, until fairly recently, have tended to be illustrative or suggestive—enough to whet one's appetite, but ultimately not very filling. In this book we attempt to show how certain parts of game theory can be applied to the rigorous development and in-depth analysis of several critical problems that afflict the security of nations, from the deterrence of foes who might launch attacks to the stabilization of crises that could explode into wars.

Three features distinguish our work from most other applications of game theory to national security:

1 A *wide range* of issues is covered. Other studies generally focus on single issues, such as deterrence, arms races, or verification.
2 The analysis is *general*. Our main interest is in analyzing superpower conflict, but the models we develop and the new concepts we define are manifestly relevant to other international conflicts as well as conflicts at lower levels.
3 The models are *deductive*. Starting from a relatively simple structure (previewed in Chapter 1), we relentlessly amend and embellish it throughout the book, inventing new models and deducing the consequences of rational play in each to create a touchstone for understanding strategic aspects of important national security issues.

The third feature deserves some amplification. We go well beyond what Roger Myerson has called "proto-game theory," or the exploitation of game theory's vocabulary and basic ideas but not its formal tools (O'Neill, 1988; a similar distinction between the descriptive and analytical uses of game theory is made in Snidal, 1985). We believe that the study of national security requires that the analyst do more than pose strategic dilemmas—as illustrated, say, by a classical 2×2 game like Prisoners' Dilemma or Chicken. To be sure, such games are our own starting point, but we add a good deal of structure to them to try to capture the essential strategic features of the situations we wish to model. Most important, we use the theory to *derive* solutions to these games that are, in some sense, optimal.

The details of our derivations are confined, in most chapters, to an appendix, which can be skipped by the nontechnical reader. In the texts of these chapters, we have endeavored to describe and fully explain our main results in relatively nontechnical language supplemented by numerous figures. (Game-theoretic and other more technical concepts used in the text are consolidated in the glossary.) Still, it may be a struggle for many political scientists, and others with little mathematical training, to follow all the steps of our arguments, in part because game-theoretic reasoning is inherently more complicated and subtle than reasoning that takes account only of the calculations of one actor.

We believe that clear strategic thinking about international relations, which by their nature are intertwined, cries out for a game-theoretic perspective. In fact, most international relations theorists would probably agree with us on the importance of this perspective; their disagreement would be on whether a formal methodology is needed to achieve this clarity of thought and hence worth the intellectual effort to master. We think, of course, that it is, but perhaps for an unexpected reason.

There are literally hundreds of contemporary books on all facets of international relations and scores of books on national security, including many on what might be phrased "U.S. national security in the nuclear age." (We know of no book with exactly this title, but it could accurately be given to several.) These books address many of the security issues addressed here, but almost always the analysis—and policy prescriptions, if any (we do not shun them in this book)—is highly qualified or imprecise, if not very confused.

True, the world is confusing, but considerable order and stability often can be found below the surface. Indeed, the objective of our equilibrium analysis is to demonstrate precisely the conditions under which we might expect stability, and what kind of stability. This search for order is the hallmark of scientific inquiry; without it, and models that highlight the order that exists, there could not be a coherent intellectual understanding of the regularities we observe.

Our results provide quantitative predictions as well as key qualitative insights. The latter are usually easier to interpret and are what we especially stress here (specific numbers mean little in our analysis). Although these insights often reinforce common sense, common sense may fail us on such a crucial issue as the *strategic* effects of Star Wars, on which there has been much heated debate but little light that transcends the rhetoric and shibboleths of proponents and opponents of SDI.

In such a controversy, we would contend, one needs to call upon rigorous models to sort out the contradictory effects and assess the net impact on deterrence—and, more generally, strategic stability—if one or both superpowers build a strategic defense system. To grasp thoroughly the implications of a carefully crafted formal model of Star Wars may, in fact, be a more economical expenditure of one's time and effort, and also more intellectually rewarding, than wading through a dozen or so books that offer many conflicting claims on the strategic effects of Star Wars but little analysis or evidence to back them up.

This brings us to our final point. A serious limitation of most of our work is that at this stage it is primarily theoretical. Although we offer a number of empirical cases as prima facie evidence of the plausibility of our models, these cases do not by themselves constitute scientific tests of the models' empirical validity.

This is probably the best that can be done at this stage, for evidence is clearly lacking: Nuclear deterrence has never failed, and Star Wars has yet to be built. Nevertheless, we hope that in the future there will be a closer marriage of game-theoretic models and relevant empirical data to corroborate more systematically some of our deductive findings or, that failing, suggest revisions in the models.

This interactive process is patently a better way to develop sound strategic knowledge than drawing conclusions inductively from a vast literature that is by turns ephemeral (specific to time or place, defying generalization), ad hoc (assertive, without theoretical foundations), or mushy (general, but vague and often inconclusive or muddled). The mushiness, especially, we try to avoid here.

Throughout the book we use masculine pronouns to describe players, but this convention is solely to simplify the discussion and avoid distraction. Emphatically, it is not intended to offend women, who are increasingly involved in national security affairs (in some cases, as heads of state).

Acknowledgments

This book is the product of two-and-a-half years of close collaboration and constant exchange between a political scientist (Brams) and a mathematician (Kilgour). To our delight, our skills proved remarkably complementary: we each speak a good part of the other's language; at a recent conference, we were even accused of thinking alike (we don't know whether this is a compliment!). Together we discussed and worked on all parts of the book, and we share responsibility for it in its entirety.

Most of the chapters originally appeared as separate papers (citations are given in the notes at the beginnings of these chapters), but they have been substantially revised for this book and their findings synthesized. We have greatly benefited from the comments of numerous colleagues and friends, often as discussants at professional meetings or as referees for the journals in which the papers were published. We will not try to acknowledge these people here but hope instead that our citations of their work in the notes provide proper acknowledgment of their contributions to this endeavor as well as their general scholarly influence on us.

We are each grateful for the financial support that we received during the research and writing of this book. Kilgour received grants from the Natural Sciences and Engineering Research Council of Canada and the Social Sciences and Humanities Research Council of Canada. He also enjoyed a sabbatical and visiting appointment at the University of Waterloo in 1984–85. Brams received grants from the National Science Foundation and the Alfred E. Sloan Foundation as well as financial assistance from the C. V. Starr Center for Applied Economics at New York University. He enjoyed a one-semester sabbatical and visiting appointment at the University of Haifa in the fall of 1984 and a Guggenheim Fellowship in 1986–87.

Our wives, Eva Brams and Joan Kilgour, met soon after we began our collaboration and have seen this book through with patience and good humor. We thank them and the rest of our families—not always so patient, but for understandable reasons—for their support and encouragement.

Figures

1

Introduction

1.1 MODELING NATIONAL SECURITY GAMES

The foundations of our analysis rest on the assumption that individuals make rational choices in decision-making situations. A thorough justification of this assumption would take us too far afield, so suffice it to say that this assumption does not connote the usual caricature of exquisitely cool, calm, and collected behavior. It simply means that individuals choose better over worse means in trying to satisfy their goals, whatever these goals are.

In *Rational Politics*, Brams has given a number of reasons to justify, and a plethora of examples to illustrate, the reasonableness of this assumption (Brams, 1985a). Although the goals of decision makers vary widely, it is argued, the rational calculations they make to try to achieve these goals pervade all levels of politics, from voting in committees to managing conflict between the superpowers.

In this volume, our concern is more specific. Positing states as players in a game, we ask: Can rational choice be imputed to decision makers in the most important strategic situations—arms races, crises that may escalate to nuclear war, and so on—that they are likely to face?

We think the answer to this question is yes, particularly if the players are nuclear powers. It is a common misunderstanding that the rationality assumption is undercut by limited information, misperceptions, restricted choices, and the like. It is not; rather, it says that *given* these constraints—some of which we build into our models—decision makers choose the best course of action they see available to them. But perhaps more to the point in the nuclear case, top decision makers, even if restricted by certain blinders, do not have to be superintelligent or able to calculate perfectly to apprehend the apocalyptic consequences of a nuclear exchange, whether it ends in a nuclear winter or some other end state slightly less dreadful.

That the horrific consequences of nuclear war are so evident, unlike the far less predictable consequences of most conventional conflicts, has been dubbed the "crystal ball effect."[1] This effect, even if the crystal ball is occasionally somewhat cloudy, reinforces the ability of decision makers to be rational by making only the crudest calculations of cost and benefit. Thereby it buttresses our case for stripping away descriptive details in nuclear conflicts in order to home in on the strategic essentials of a situation.

Doubtless, our austere representations of these situations as formal games will appear to many to be quite unrealistic. But if the players are nuclear powers, their choices often *are* stark. Since the parsimonious representation of these choices in games is precisely what enables us to explore their implications rigorously and in depth, the austerity seems both necessary and desirable.

The advantages of parsimony and rigor are considerable. In particular, they enable us to ascertain *all* possible stable outcomes in each game studied, specify those (if any) that lead to desired outcomes, and even chart economical ways to induce their choices. In addition, the spare theoretical structures that we start with allow us to add new rules or constraints without making the resulting games analytically unmanageable. These new games, moreover, often raise entirely new strategic questions that we are then able to pursue.

Before explicating the concept of rationality in the games we will analyze, let us deal with the simplification that nation-states can be treated as single players. Manifestly, no reasonably complete description of international behavior can consider nation-states to be unitary actors, with the possible exception of states run by dictators with absolute control. Because there are few if any such autocrats today, our simplifying assumption would appear dubious if not untenable. The unitary-actor assumption is nonetheless useful in explaining international behavior. Furthermore, except for a very few chaotic countries, like Lebanon today, it usually is realistic in the sorts of serious conflicts with other states that we model in the following chapters. Generally speaking, if there is no consensus among top decision makers, they proceed to formulate some policy on which agreement can be reached. In such situations, the state may be considered to act *as if* it were a unitary actor.

The consensus may break down after the conflict, especially if the policy was unsuccessful and there are then recriminations against the state's leaders. But in most major confrontations or crises, nation-states can be considered as unitary actors for the purpose of modeling their *international* behavior.

Of course, domestic political games among elites are ubiquitous. However, these almost always pale to insignificance when issues of war and peace come to the fore and a nation-state, its national security imperiled, must act. Its key leaders do act, and usually together, making the unitary-actor assumption a sensible theoretical simplification in matters involving national security.

By *national security* we refer to those circumstances or events that directly affect the safety or integrity of states in their relations with other states. For example, whether a state's economy prospers or declines does not usually affect its survival in the international system, though a precipitous economic decline that turns into a deep depression may so weaken a state that it becomes vulnerable to exploitation by its enemies. Then we would say its national security is jeopardized. However, the usual threat to a state's vital national interests comes not from economic decline but from war (the two, of course, may be related).

The games we will analyze are not war games as such, but the choices that players make may precipitate conflict that leads to war. For example, some arms races, analyzed in chapters 2 and 4, culminate in war, and the breakdown of deterrence, analyzed in chapters 3 and 4, can have devastating consequences, particularly if the countries involved are nuclear powers.

Indeed, the use of nuclear weapons by one country against another, as has been threatened on several occasions in the past,[2] could affect international security above and beyond the national security of the states immediately involved. Even the limited use of nuclear weapons could destroy the foundations of the present international order.

In addition to ensuring their own national survival, it is manifestly in the interest of states to preserve some international order. As we shall see, the national security games nations play, especially those between the superpowers, give *everybody* a stake in the outcome.

This interdependence would appear to cast doubt on our assumption, in all the games analyzed, of only two players. We justify this assumption on grounds similar to those on which our unitary-actor assumption rests. It is useful, and certainly not implausible, in many international conflicts. To be sure, such conflicts often involve more than two states, as have all five major Arab-Israeli wars (1948, 1956, 1967, 1973, and 1982) in varying degrees.

Because the Arab states generally cooperated in their actions against Israel, however, they can properly be treated as a single player. Although the Arab world has always been rife with internal conflict, in national security matters involving Israel these differences have usually been submerged or patched over, at least temporarily. Even Israel's peace treaty with Egypt, though it precludes war, has not engendered particularly warm or friendly relations between these countries lately.

Another consideration pertinent to fixing the number of players in a game-theoretic model is the power—difficult as this concept is to define—that they can exercise. Although each interest group may be treated as a player, some may properly be dropped from the analysis because they are "dummies"—that is, they have no power to affect the outcome. In all our two-person models the actors are presumed to be genuine players who can affect the course of the conflict. Our exclusion of other actors does not signal that they have no interest in the outcome but rather that they have no significant influence on its choice.

In developing and applying our models, the principal real-world conflict we seek to understand and explain is that between the superpowers. In this sense, our book can be considered a sequel to Brams's *Superpower Games* (Brams, 1985b), for we have tried to extend much of the analysis in that book to more complex strategic situations involving quantitative, sequential choices. In addition, we investigate entirely new games within the general framework of rational choice.

Our analysis can be thought of as adding shades of gray to the black-and-white choices of earlier models. Thus, instead of the dichotomous options of cooperation and noncooperation, we allow each player to select continuous *levels* of cooperation (or noncooperation). We also allow for the possibility of retaliation at any level if the initial choice of an opponent was regarded either as noncooperative or as more noncooperative than one's own initial choice.

This basic structure characterizes two games based on Prisoners' Dilemma and four games based on Chicken (with the games connected by the arrows being analogues of each other—they are played according to the same set of rules):

Prisoners' Dilemma   Chicken

Deescalation Game (chapter 2)   ⟷ Deterrence Game (chapter 3)
Arms Reduction Game (chapter 4) ⟷ Winding-Down Game (chapter 4)
                                   Star Wars Game (chapter 5)
                                   Threat Game (chapters 6 and 7)

In fact, the Star Wars and Threat Games also have Prisoners' Dilemma analogues, but we do not analyze them here. The reason we do not is that the questions the Star Wars and Threat Games raise about strategic defense, optimal threats, and crisis stability seem best understood through Chicken, on which they are based.

It is worth noting that the Threat Game, while allowing the players continuous, sequential choices, differs significantly in structure from the other games listed. In the Threat Game, each player can tailor his threat of retaliation to the initial level of preemption of his opponent. A convenient interpretation of the choices of the players in the Threat Game is the following: Together they move a marker on a game board; one player controls its movement in the horizontal direction and the other its movement in the vertical direction.

The principal solution concept studied in the Threat Game, as in the other games, is an equilibrium notion proposed by Nash (1951). We will illustrate it shortly when we discuss the meaning of rational play in games, but first some other characteristics of these games deserve to be mentioned.

Besides being two-person, continuous, sequential games, they are "noncooperative" and "variable-sum." *Noncooperative* means that the players cannot make binding or enforceable agreements, and variable-sum means roughly that they both can "win" or both can "lose" simultaneously (a more precise definition will be given in chapter 2), vitiating the usual meaning of win and lose. For example, both players may be hurt if there is a costly arms race, whereas both may benefit if they can reach an arms control agreement.

The games analyzed here are all noncooperative, however, which means that such an agreement is assumed to be unenforceable unless, of course, the players themselves choose to adhere to it. Hence, as will be discussed presently, the players' temptation to abrogate an agreement is crucial to our analysis of arms control and verification.

The final game we will analyze is the Verification Game (chapter 8), which is not, like the other games, *symmetric*: The players, an inspector and an inspectee, face different strategic choices, whereas in the other games the players' options are strategically equivalent (this equivalence will be defined precisely later). Also, besides Nash equilibria, we analyze certain outcomes in the Verification Game that the inspector can induce the inspectee to choose by virtue of the fact that the inspector can detect treaty violations with a known probability and take appropriate countermeasures based on what he detects.

Since the inspector is assumed to have only an imperfect detector that can be thought of as depending on a chance device, the Verification Game is technically one of "imperfect information." Similarly, none of our other models is based on a game of *perfect information*, wherein each player knows the strategy choice of his opponent at each stage of play. Because these other games involve simultaneous and hence independent choices, neither player can act on knowledge of exactly what choice his opponent makes at these stages.

All the games analyzed in these chapters are games of *complete information*. This means that the players are fully informed of the rules of play; in fact, a *game* can be defined to be the sum total of its rules of play. These rules describe, among other things, the payoffs to the players at every outcome, which are rooted in how much the players prefer each outcome over the others.

So much for the games that we investigate in later chapters. The game *theory* we apply to these games is designed to illuminate the calculations of players who seek to maximize their "expected payoffs" (defined and illustrated in section 2.3), taking into account that an opponent is making similar calculations.

These interdependent calculations differ from those made in *decision theory*, in which "states of nature" are assumed to arise exogenously according to a probability distribution, making such a model a one-person "game against nature." Whereas nature is assumed to be indifferent or neutral in choosing states in decision theory, players in game theory have preferences, often conflicting, and select strategies accordingly (illustrated in section 1.2). Furthermore, outcomes depend on the choices of *all* players, making players' decisions truly interdependent. Which outcomes are stable, and what utility players derive from them, are the principal questions we address in subsequent chapters.

We next turn to a brief description and analysis of the structure underlying all our games except the Verification Game. The threat structures we incorporate in these other games can be represented (more or less) by a generic two-person game we call the Conflict Game.

1.2 THE CONFLICT GAME

In the Conflict Game there are two stages: At the *first stage*, each player can choose either to cooperate (C) or not cooperate ($\bar{C}$); at the *second stage*, each player who chose C can choose either to retaliate (R) or not retaliate ($\bar{R}$) if his opponent chose $\bar{C}$ in the first stage (of which he is informed).[3] If both parties choose $\bar{C}$ at the first stage, we assume that the game does not proceed to the second stage: Having already chosen noncooperation (e.g., escalation in an arms race, presumption in a crisis), the players will be unable to heighten the conflict further through retaliation.

On the other hand, if one player chooses C while his opponent chooses $\bar{C}$ in the first stage, the C-player's choice of R or $\bar{R}$ comes into play. Consider the following two plans for the C-player on how to act at this stage, contingent on his opponent's first-stage choice:

1 Choose R at the second stage if his opponent chose $\bar{C}$ at the first stage, $\bar{R}$ if his opponent chose C (tit-for-tat).
2 Choose $\bar{R}$ regardless of what his opponent did at the first stage (unconditional cooperation).

Coupling these second-stage plans with the choice of C at the first stage gives a player two *strategies*, or complete plans on how to act at each stage: C and tit-for-tat, which we call CR; and C and unconditional cooperation, called $C\bar{R}$. In sum, these two strategies and strategy $\bar{C}$ (i.e., $\bar{C}$ at the first stage, which terminates future choices) give each player three strategies in the Conflict Game: $\bar{C}$, CR, and $C\bar{R}$.

These strategies for players A and B are shown as the rows (A) and columns (B) in the *payoff matrix* of figure 1.1. A matrix like this one is called the *normal-form* or *strategic* representation of a game, wherein each possible outcome—and the payoffs to the players associated with it—corresponds to a cell at the intersection of a pair of strategies of the two players. Since each player has three strategies, the 3×3 payoff matrix of figure 1.1 depicts the nine possible strategy combinations and associated payoffs in the Conflict Game. The payoffs are simply the *utilities*, or numerical values, the players attach to the outcomes.

B

	$\bar{C}$	CR	C$\bar{R}$
$\bar{C}$	TR $[(r_T,c_T)]$	BR (r_B,c_B)	AW $[(r_4,c_2)]$
A CR	AR (r_A,c_A)	SQ $((r_3,c_3))$	SQ (r_3,c_3)
C$\bar{R}$	BW $[(r_2,c_4)]$	SQ (r_3,c_3)	SQ (r_3,c_3)

Key: (r_i,c_j) = (payoff to A, payoff to B)
Circled outcome is *always* a Nash equilibrium.
Boxed outcomes are *possible* Nash equilibria.
C/$\bar{C}$ = cooperate/don't cooperate; R/$\bar{R}$ = retaliate/don't retaliate

Figure 1.1   Payoff matrix of Conflict Game.

There are six *distinct* outcomes in this game, which we characterize by a verbal description (and abbreviation), a payoff (utility) for A, and a payoff (utility) for B:

Outcomes	Payoff for A	Payoff for B
Status quo (SQ)	r_3	c_3
A wins (AW)	r_4	c_2
B wins (BW)	r_2	c_4
A retaliates (AR)	r_A	c_A
B retaliates (BR)	r_B	c_B
Trap (TR)	r_T	c_T

In the figure 1.1 payoff matrix, the payoffs to the players at each outcome are represented by ordered pairs (r_i,c_j), where r_i is the payoff to the row player (A) and c_j the payoff to the column player (B).

We make the following assumptions about the relationships among each player's payoffs:

1 *Each player prefers winning to the status quo*: $r_4 > r_3$ and $c_4 > c_3$
2 *Each player prefers the status quo to the other player's winning*: $r_3 > r_2$ and $c_3 > c_2$

3 *Each player prefers the status quo to any of the three non-cooperative outcomes*: $r_3 > \max\{r_A, r_B, r_T\}$ and $c_3 > \max\{c_A, c_B, c_T\}$

Because the function "max" (maximum) chooses the largest quantity in the set it precedes, assumption 3 ensures that r_3 is greater than all three r_i's with letters as subscripts, and similarly for c_3. Note that the greater the subscript for each r_i and c_j with numerical subscripts, the greater its value to A and B, respectively.

Assumptions 1–3 are sufficiently unrestrictive that the Conflict Game can subsume both Prisoners' Dilemma (defined in chapter 2) and Chicken (defined in chapter 3) when their rules are modified to allow for possible retaliation by one or both players in a second stage, given that both players do not choose $\bar{C}$ in the first stage and thereby cut off a future choice of R or $\bar{R}$. To identify whether the Conflict Game is Prisoners' Dilemma or Chicken requires that the preferences of the players for the three noncooperative outcomes—AR, BR, and TR—be specified.

To begin our analysis of the Conflict Game, we define an outcome to be *Pareto-superior* if there exists no other outcome that is better for one player and at least as good for the other. By this definition, three of the six outcomes are Pareto-superior: SQ, yielding (r_3, c_3); AW, yielding (r_4, c_2); and BW, yielding (r_2, c_4). The other three distinct outcomes are *Pareto-inferior* because, by assumption 3, (r_3, c_3) is better for both players than any of these noncooperative outcomes. This fact gives each player the leverage to *threaten* retaliation (R) if his opponent is noncooperative ($\bar{C}$) at the first stage—by choosing strategy CR—in order to induce him to be cooperative at this stage. Such a threat, if credible, can stabilize the status quo (SQ), which may be regarded as a kind of compromise outcome for both players: For each player, it is inferior to winning (at AW or BW) but superior to the other player's winning (at BW or AW) and to any of the noncooperative outcomes (AR, BR, or TR).

Of the four SQ strategy combinations, only the one circled SQ in figure 1.1 is *always* a *Nash equilibrium*: Neither player would ever have an incentive to depart unilaterally from his CR strategy associated with this outcome, because he would do worse, or at least not better, if he did. Thus, if A switched to strategy $\bar{C}$, he would get a payoff of r_B (worse), and if he switched to C$\bar{R}$ he would get r_3 (no better). Hence, he would have no incentive to deviate from strategy CR, associated with this SQ outcome, given that his opponent did not switch from his CR strategy; for analogous reasons neither would B have an incentive to deviate from his CR strategy.

By contrast, the other three cells can never be Nash equilibria, for at least one player can improve on each of these outcomes by departing unilaterally [either A from the two (r_3, c_3) outcomes in the third column to (r_4, c_2), or B from the two (r_3, c_3) outcomes in the third row to (r_2,c_4)]. In fact, it is the instability of these compromise outcomes in the *absence* of retaliatory threats by both players that motivated us to search for precise conditions that would render such outcomes Nash equilibria in the various games to be analyzed in later chapters.

Unfortunately, the apparent rationality of the choice of the circled SQ outcome in figure 1.1, though always a Pareto-superior Nash equilibrium, may be upset if

$$r_2 > r_A \tag{1.1a}$$

or

$$c_2 > c_B \tag{1.1b}$$

Consider (1.1a). If it is satisfied, A's strategy $C\overline{R}$ *dominates* CR: It is at least as good (when B chooses CR or $C\overline{R}$), and in one contingency better (when B chooses $\overline{C}$), than CR. Similarly, B's strategy $C\overline{R}$ dominates CR when (1.1b) is satisfied.

As a consequence, unless both inequalities of (1.1) fail, it would appear that the players would *not* be well advised to choose their strategies associated with the (r_3,c_3) Nash equilibrium.[4] For in this circumstance, the CR strategy of at least one player wil be *dominated*: Every outcome it leads to is never better, and can be worse, than the corresponding outcome given by his $C\overline{R}$ strategy. In such a case, a player would, presumably, choose his $C\overline{R}$ strategy over his dominated CR strategy.

The problem with the switch to $C\overline{R}$ by one player is that it results in an (r_3,c_3) outcome that is *never* in equilibrium. In fact, as soon as a player's choice of $C\overline{R}$ is anticipated by an opponent, the opponent would have good reason not to choose CR or $C\overline{R}$ but instead $\overline{C}$, leading to his best outcome (r_4 or c_4) and an inferior outcome (c_2 or r_2) for the player who switched originally from CR to $C\overline{R}$.

By this reasoning, $C\overline{R}$ would seem an inopportune choice, even when it dominates CR. More opportune for a player in such a situation would be to stick with CR, for he loses nothing over the choice of $C\overline{R}$ if his opponent also chooses CR: the same (r_3,c_3) payoff would be realized, and it is, of course, stable. Note that if one or both of the (1.1) inequalities holds, the boxed outcomes in figure 1.1 may also be Nash equilibria. Among these, (r_T,c_T) is, by assumption 3, Pareto-inferior to the (r_3,c_3) Nash equilibrium and hence an outcome it is in the best interest of both players to avoid.

In order to strengthen the choice of the (r_3,c_3) Nash equilibrium, players in the Conflict Game might not only claim that they are irrevocably *precommitted* to retaliation but also expend considerable effort to make this claim credible. (In chapter 3 we argue that the superpowers in fact do so in the Deterrence Game, which is a specific form the Conflict Game takes when additional structure is incorporated into it.) To see how precommitments work, notice that even if B believes that $r_2 > r_A$, he will still be predisposed to choose CR if he believes that A is committed to choosing CR. Indeed, if B knows that A will choose CR, then he can do no better than choose CR himself, which follows from the fact that the resulting outcome is a Nash equilibrium.

If the players in the Conflict Game manage to convince themselves that retaliation is certain, then the game is actually different from that shown in figure 1.1. Specifically, the outcomes AR and BR, in which one player wins, become merely transitory. Should they—and a $\overline{CR}$ strategy—be momentarily selected, certain retaliation by the losing (cooperative) player implies that he will immediately switch to CR.

Put another way, if a player chooses $\overline{CR}$, his commitment to retaliation means that he will not stick with it if attacked but will instead switch to CR. Consequently, both $\overline{CR}$ strategies can be deleted from the 3×3 Conflict Game, once retaliation is certain, reducing it to a 2×2 game. In this smaller game, there is only one SQ outcome; moreover, certain retaliation wipes out the distinction among the three noncooperative outcomes—AR, BR, and TR—rendering them all TR. Because SQ is the unique Pareto-superior Nash equilibrium in this 2×2 game, it will presumably be the rational choice of the players.

As we show in subsequent chapters, building the threat of retaliation into the second stage of different conflicts will, under varying conditions, induce cooperative choices at the first stage. Furthermore, games that the 3×3 Conflict Game adumbrates, which reflect second-stage choices and permit retaliatory threats, can generally be reduced to 2×2 games, as we have demonstrated here. This reduction is assumed in the remainder of the book.

The alternative to persuading an opponent that aggressive acts will be punished by subsequent retaliation is to persuade him that the appropriate inequality in (1.1) does not hold. In some conflicts, this may be true: Retaliating, with some prospect of turning the tide, will be better than suffering ignominious defeat.

However, there are other conflicts, such as a nuclear confrontation between the superpowers, in which the truth of inequalities (1.1) is not open to doubt: "Losing" may well be better than allowing a conflict to spiral out of control, risking Armageddon. In such conflicts, a player cannot credibly maintain that he will retaliate against any aggression unless he precommits his forces to respond in a manner somewhat beyond his control. As we shall see in chapter 3, the retaliation procedures that have been instituted by both superpowers for responding to a first strike are really attempts—happily, so far successful—to buttress nuclear deterrence by making one's precommitment to carrying out a second strike as believable as possible.

Still, it is hard to imagine how rational players could become absolutely convinced that threatened retaliation will actually occur when it is evident that it would harm the threatener as well as the threatened party. Fortunately, such absolute conviction is not necessary. The SQ outcome associated with the CR strategies of both players will still be in equilibrium, even if (1.1) fails, provided that both players think their opponent's retaliation is likely enough.

In the appendix to this chapter, we drive this weaker requirement and in so doing show how expected payoffs can be defined by refining our analysis of the Conflict Game. We do this by allowing the players to choose *mixed strategies*, which are probability distributions over their *pure strategies* (i.e., specific strategies, such as those shown in figure 1.1, chosen with certainty).

Thereby we demonstrate that it is not necessary for each player to believe in the certainty of his opponent's retaliation for the SQ equilibrium to be a compelling solution to the Conflict Game. Rather, it is sufficient that each player believe that the probability that his opponent will retaliate exceeds a certain threshold. Thus, an uncertain but frightful enough threat, by stabilizing the SQ outcome in the Conflict Game, can suffice to keep enemies at bay.

1.3 CONCLUSIONS

We called the Conflict Game "generic" because we left unspecified certain relationships among the outcomes. At this early point in the book, we simply wished to outline the problem that two players face—starting out from a cooperative status quo outcome—when they desire to stay at this outcome but, at the same time, there are obvious incentives to depart from it in order to try to do better. (In later chapters we also reverse this process and examine how players can escape from a conflictual to a cooperative outcome via a path that may entail some risk.) To preserve cooperation in the Conflict Game requires threats of retaliation, but they need not be carried out with certainty to deter an opponent who may become aggressive.

In later chapters we become much more specific about the nature and level of these threats as well as their interpretation in different kinds of conflicts. But if the incorporation of threats into the structure of games is the keynote of this work, it is not the only theme. We introduce additional parameters into our games and, in the case of the Threat Game, alter the rules of play of the Conflict Game in order to analyze features of national security conflicts that require different rules to model. In the Verification Game, we go one step further and limit the inspector's ability to counter cheating by his inability always to detect it, just as the players' strategic defenses in the Star Wars

Game may constrain their ability to preempt an opponent or retailiate against such preemption.

To paraphrase an old saying, national security is too important to leave to game theorists or political scientists, especially those who build abstract models. Yet if game theory does not provide an immediate solution to the next crisis, it does, we believe, help greatly in thinking carefully about national security policy, especially its seeming paradoxes (e.g., those connected with nuclear deterrence).

We are not reticent about drawing policy lessons from our models, but we generally refrain from prescribing what should be done about such-and-such in the here-and-now. Instead, we try to identify (1) fundamental problems in preserving peace and (2) rational approaches to solving or alleviating these problems. In the final chapter, in particular, we consider briefly how the probability of nuclear war may be reduced over the long run by acting upon some of the findings of our strategic analysis.

It is worth pointing out that, for all the strategic issues taken up in this book, there are a number of important areas we ignore. We have already noted that our models are restricted to two players, precluding the analysis of coalitions. We therefore have little to say about the subject of alliances, including their cohesion and effectiveness under different circumstances, or how deterrence between the superpowers impinges on allies ("extended deterrence"). Balance of power, an influential if controversial concept in international relations, is also neglected because it presumes multiple actors, some of whom seek to prevent the hegemony of a single player.[5] Finally, bargaining and negotiation mechanisms and strategies have received much attention from game theorists recently, but there are few significant applications of these models to international relations.[6]

These are but a few of the gaps that deserve the attention of formal theorists who wish to apply their skills to the analysis of critical problems of national security. Our knowledge of these problems, not to mention possible solutions, is still inchoate; the amorphous, fuzzy, and speculative thinking that abounds in the national security field has not helped matters. We believe that the development of game-theoretic models that address specific questions, including the hard trade-offs of one kind of security for another that may be necessary, can give a substantial boost to systematic strategic analysis.

Although the intellectual benefits that game-theoretic models can provide are considerable, they have a price: The serious study of strategic issues is not easy. These models require prolonged contemplation even to formulate well, not to mention the searching analysis required to ensure that their conclusions are cogent and illuminate a significant problem. Nevertheless, this price seems well worth paying on matters of such consequence to our survival.

APPENDIX

We begin by developing notation to express players' mixed strategies, or probability distributions over pure strategies, in the Conflict Game of figure 1.1. Suppose that player A has chosen a probability distribution over this three pure strategies such that their probabilities are $P_A\{\bar{C}\}$, $P_A\{CR\}$, and $P_A\{C\bar{R}\}$, where $P_A\{\bar{C}\} + P_A\{CR\} + P_A\{C\bar{R}\} = 1$. Define s, the probability that A will cooperate initially:

$$s = P_A\{CR\} + P_A\{C\bar{R}\} = 1 - P_A\{\bar{C}\}$$

If $s > 0$, define q, the probability that A will not retaliate if preempted:

$$q = \frac{P_A\{C\bar{R}\}}{P_A\{CR\} + P_A\{C\bar{R}\}}$$

(If $s = 0$, q is arbitrary.) Note that $s = P_A\{C\}$, the probability that A cooperates initially, and $q = P_A\{\bar{R} \mid C\}$, the conditional probability that A will not retaliate if B is noncooperative initially, given that A is cooperative initially. Our notation has been chosen to be consistent with later chapters, where cooperative behavior is emphasized. Thus s and q are both probabilities of acting cooperatively.

Analogously, we represent player B's mixed strategy by two quantities, t and p, that denote probabilities of cooperative behavior by B:

$$t = P_B\{CR\} + P_B\{C\bar{R}\} = 1 - P_B\{\bar{C}\} = P_B\{C\}$$

$$p = \frac{P_B\{C\bar{R}\}}{P_B\{CR\} + P_B\{C\bar{R}\}} = P_B\{\bar{R} \mid C\}$$

As before, p is defined only if $t > 0$; when $t = 0$, p is arbitrary.

It is not difficult to show that if A's strategy is (s, q) and B's is (t, p), then the players' expected payoffs (expected utilities) in the Conflict Game are

$$E_A(s,q;t,p) = str_3 + s(1 - t)\,[qr_2 + (1 - q)r_A] \\ + (1 - s)t\,[pr_4 + (1 - p)r_B] + (1 - s)(1 - t)r_T \quad (1)$$

$$E_B(t,p;s,q) = stc_3 + (1 - s)t[pc_2 + (1 - p)c_B] \\ + s(1 - t)\,[qc_4 + (1 - q)c_A] + (1 - s)(1 - t)c_T \quad (2)$$

These calculations will be explained in detail for specific games later.

From the figure 1.1 matrix and the strategy definitions above, it is clear that a status quo (SQ) outcome will be chosen with certainty iff (if and only if) $s = 1$ and $t = 1$. We now use (1) and (2) to derive precise conditions under which there is a Nash equilibrium with $s = 1$ and $t = 1$. Although we presume assumptions 1–3 in the text, we do not assume inequalities (1.1) in what follows.

If $t = 1$, substitution in (1) gives

$$E_A(s,q;1,p) = [pr_4 + (1 - p)r_B] + [(r_3 - r_B) - p(r_4 - r_B)]s \qquad (3)$$

Now A maximizes $E_A(s,q;1,p)$ by choosing $s = 1$ iff the coefficient of s on the right side of (3) is non-negative, which occurs iff

$$p \leqq \frac{r_3 - r_B}{r_4 - r_B} \qquad (4)$$

Parenthetically, we note that assumptions 1 and 3 in the text imply that

$$0 < \frac{r_3 - r_B}{r_4 - r_B}$$

so that condition (4) is well-defined.

Considering how B maximizes his expected payoff by choosing $t = 1$ in the face of A's choice of $s = 1$ gives a condition analogous to (4):

$$q \leqq \frac{c_3 - c_A}{c_4 - c_A} \qquad (5)$$

[Condition (5) is similarly well-defined.] It is easy to verify that the conjunction of the two necessary conditions, (4) and (5), is sufficient for an equilibrium.

We conclude that there is a Nash equilibrium with $s = 1$ and $t = 1$ iff (4) and (5) hold. This yields an equilibrium pair of strategies with outcome SQ (for certain): No player ever chooses $\bar{C}$. However, each player might retaliate if the other were to choose $\bar{C}$; A would retaliate with a probability at least $(r_4 - r_3)/(r_4 - r_B)$, and B with a probability at least $(c_4 - c_3)/(c_4 - c_A)$.

NOTES

1 Harvard Nuclear Study Group (1983, pp. 43–44). In assessing the various incentives and disincentives to use nuclear weapons in possible future conflicts, Nacht (1985) concludes that the disincentives are enormous but nevertheless need to be strengthened to ensure still greater stability. See also Lebow (1987).

That an "ultimate deterrent" is already in place and superpower stability is secure is argued by Shepherd (1986); related arguments that the world is freer of the risk of nuclear war are given by Snow (1987).

2 See Bundy (1986, p. 41); Halloran (1986); *New York Times*, July 22, 1985, p. A12; Mendolovitz (1985, pp. 39–40); and Betts (1987).

3 Multistage choices and sequential play, not unlike that developed here for the Conflict Game, lie at the heart of Wagner's analysis of several games (Wagner, 1982, 1983). This approach is also taken in Witt (1986). More generally, see O'Flaherty (1985), which introduces the possibility of making commitments, comprising both promises and threats, into the definition of equilibrium strategies.

4 In another formulation, if either of these inequalities holds, this equilibrium is "imperfect"; in chapter 3 we consider this concept further and discuss its implications.

5 For recent game-theoretic models in this area, see Wagner (1986) and Niou and Ordeshook (1986, 1987).

6 A good review of this literature at a relatively nontechnical level, including applications to international relations (negotiation of the Panama Canal Treaty, the Camp David agreement, and the Law of the Sea Treaty), can be found in Raiffa (1982). At a more technical level, see Roth (1985).

2

Arms Races

2.1 INTRODUCTION

The prevention of nuclear war is surely the most daunting problem facing the world today. The road to such a war, should one ever occur, will probably not be a "bolt from the blue"—a massive nuclear surprise attack by one superpower against the other and its allies. Rather, it is likely to erupt in a period of extreme crisis occasioned by a conventional conflict in which one side, facing imminent defeat, decides it has no recourse except to use nuclear weapons or threaten their use. The conflict need not even involve a nuclear power directly but only an ally that feels compelled to come to the aid of a threatened partner.

An arms race may trigger such a conflict. As tensions mount in such a race, verbal threats and provocative military maneuvers may precipitate war, which may well escalate as allies become involved. Then, if one side's position or very existence is jeopardized, there is a possibility that it would introduce or threaten to introduce nuclear weapons to try to avert disaster.

In subsequent chapters, we analyze what kinds of probabilistic threats seem to be optimal in preventing confrontation situations (modeled by the game of Chicken) from exploding and wreaking destruction on both sides. In this chapter, we focus on the progenitor of many crises that produce such perilous showdowns—namely, arms races. Our aim is to show under what conditions deescalation rather than escalation is a rational response to the staggering burdens that an unrestrained arms race imposes on both sides.

For this purpose, we start from a model of an arms race based on the infamous game of Prisoners' Dilemma (to be described in section 2.2), but we make major emendations in the simple 2×2 version of this game to permit each player

1 *Initially* to choose any level of provocation along a disarm-arm dimension
2 *Subsequently* to choose any level of response to a provocation if it

is viewed as escalatory or noncooperative, provided the player's own initial choice was considered cooperative

The levels of provocation, and response to provocation, of the players are assumed to correspond to *probabilities* of escalation, and retaliation for escalation, that will be defined in section 2.3.

We assume that each player chooses a probability of escalation and a probability of retaliation for escalation from infinite strategy spaces. These two probabilities, each of which can be any number between 0 and 1 (inclusive), are selected at the beginning of play. The escalation probabilities determine which of the four outcomes of Prisoners' Dilemma occurs initially. If the initial outcome is one in which one player escalates and the other does not, the nonescalatory player's retaliatory probability comes into play and may move the game to the noncooperative outcome of Prisoners' Dilemma. It should be emphasized that the resulting two-stage game, which we call the *Deescalation Game*, is no longer Prisoners' Dilemma.

After calculating *maximin strategies* (which maximize each player's minimum payoff) in the Deescalation Game, we demonstrate that it contains two Nash equilibria, or stable outcomes. The one we call the *escalation equilibrium* corresponds to the unique Nash equilibrium in the classical 2×2 Prisoners' Dilemma game. The other, which we call the *deescalation equilibrium*, involves each side's cooperating initially with certainty but retaliating with a specified probability to noncooperation by the other side. Although the deescalation equilibrium is a promising addition to the classical version of Prisoners' Dilemma (without the possibility of retaliating), its existence does not answer the nagging question of how one extricates onself from the escalation equilibrium of the Deescalation Game, which by definition neither player has an incentive to depart from unilaterally.

The superpowers seem stuck at this noncooperative equilibrium today. Felicitously for the players in the Deescalation Game, however, there is a trajectory, or path, by which they can travel from the escalation equilibrium to the deescalation equilibrium. Surprisingly, as we shall show, either player can initiate such a sequence with impunity, triggering subsequent rational moves by the players that redound to the benefit to both, eventually reaching the deescalation equilibrium.

We will briefly compare this resolution of the trying dilemma posed by arms races—particularly that between the superpowers—to other game-theoretic approaches. We believe that our model offers a more realistic representation of the superpower arms race than others, some of which, nonetheless, suggest a resolution similar to ours.

2.2 ARMS RACES AND PRISONERS' DILEMMA

The 2×2 game of Prisoners' Dilemma, in which the two players (called
Row and Column) each have two strategies and can rank the resulting
four outcomes from best (4) to worst (1), is illustrated in figure 2.1.
The first number in the ordered pair that specifies each outcome is
assumed to be the ranking of Row, and the second number the ranking
of Column; the higher the ranking, the more preferred the outcome
for each player.

		Column	
		Cooperate (C)	Do not cooperate ($\bar{C}$)
	Cooperate (C)	(3,3) Compromise	(1,4) Column wins
Row	Do not cooperate ($\bar{C}$)	(4,1) Row wins	(2,2) Trap

Key: (x, y) = (rank of Row, rank of Column)
   4 = best; 3 = next best; 2 = next worst; 1 = worst
   Circled outcome is Nash equilibrium.

Figure 2.1   Outcome matrix of Prisoners' Dilemma.

Thus, for example, the outcome (3,3) is next best for both players.
However, we make no presumption about whether this outcome is
closer to each player's best (4) or next-worst (2) outcome. (Later we
assume that players can assign numerical values, or cardinal utilities, to
the outcomes.) Because the two players do not rank or order any two
outcomes the same—that is, there are no equally preferred outcomes
—this is a *strict ordinal game*.

The shorthand verbal descriptions given in figure 2.1 for the
outcomes are intended to convey the qualitative nature of the
outcomes, based on the players' rankings. Because this game is
symmetric (i.e., the players rank the two outcomes along the main
diagonal the same, and the ranks of the off-diagonal outcomes are
mirror images of each other), both players face the same problem of
strategic choice.

Each player is assumed to be able to choose between the strategies
of cooperation (C) and noncooperation ($\bar{C}$). Each obtains his next-best
outcome of 3 ("compromise") by choosing C if the other player also
does, but both have an incentive to defect from this outcome to obtain
their best outcomes of 4 by choosing $\bar{C}$ when the other player chooses
C. Yet if both choose $\bar{C}$, they bring upon themselves their next-worst

outcome ("trap"). On the other hand, should one player choose $\bar{C}$ when the other chooses C, the $\bar{C}$-player wins by obtaining his best outcome (4) at the same time that the C-player suffers his worst (1) outcome.

The dilemma in Prisoners' Dilemma is that both players have a *dominant* strategy of choosing $\bar{C}$: Whether one's opponent chooses C or $\bar{C}$, $\bar{C}$ is a better choice. But the choice of $\bar{C}$ by both leads to (2,2), which is Pareto-inferior since it is worse for both players than (3,3). In addition, (2,2) is a Nash equilibrium, whereas (3,3) is not stable in this sense.

Thus, for example, Row could do immediately better by switching from C to $\bar{C}$, moving the outcome from (3,3) to (4,1), and similarly Column could do better moving from (3,3) to (1,4). From (2,2), however, a unilateral switch of strategy by either player would lead to his worst outcome, thereby rendering (2,2) stable.

Presumably, rational players would choose their dominant, or unconditionally best, strategies of $\bar{C}$, leading to the Pareto-inferior (2,2) Nash equilibrium. Because of its stability, neither player would be motivated to depart from (2,2), even though (3,3) is a better outcome for both than (2,2). In fact (3,3) is Pareto-superior since any other outcome that is better for one player is worse for the other [outcomes (4,1) and (1,4) are also Pareto-superior]. Should (3,3) somehow be chosen, however, both players would be tempted to depart from it to try to do still better, which is what renders it unstable.

Other concepts of equilibrium distinguish (3,3) as a stable outcome, but the rules of play they assume require that players act nonmyopically or farsightedly; moreover, they do not rule out (2,2) as stable also (see Brams and Wittman, 1981; Kilgour, 1984; Zagare, 1984). If threats are possible in repeated play of Prisoners' Dilemma under still different rules, the stability of (3,3) is reinforced (Brams and Hessel, 1984). Preplay negotiations can also lead to the (3,3) outcome (Kalai, 1981).

We will introduce shortly the notion of a probabilistic threat as well as a probabilistic initial strategy choice. But before doing that, it is worth pointing out that Prisoners' Dilemma is not a *constant-sum game*, in which what one player wins the other player loses.[1] Rather, it is a *variable-sum game*, because the sum of the players' payoffs at each outcome (if measured cardinally by utilities rather than ordinally by ranks) may vary.

A variable-sum game is also a *game of partial conflict*, as opposed to a (constant-sum) *game of total conflict* in which the players' preferences are diametrically opposed: In the latter, one player cannot benefit except at the expense of another. Prisoners' Dilemma is not a game of total conflict, for both players do worse at (2,2) than at (3,3);

perhaps this fact belies the name "partial conflict," since (2,2) is, unfortunately for the players, both the product of dominant strategies and the unique Nash equilibrium. It is hard to see how the players can avoid it without risking their worst outcomes.

As a model of the superpower arms race, this recalcitrant game supports the logic of both sides' arming (noncooperation), even though this outcome is Pareto-inferior to their disarming or, less ambitiously, pursuing more limited policies of arms control (cooperating). Cooperation is problematic because, as Garthoff puts it, "they [the Soviets] would like to have an edge over us [at (1,4) if they are Column], just as we would like to have an edge over them [at (4,1) if we are Row]."[2]

Prisoners' Dilemma elegantly captures this temptation to defect from the cooperative outcome that, it seems, has inexorably led the superpowers into a very costly arms race. Nevertheless, at the same time that it offers a striking explanation of the fundamental intractability of this continuing conflict—based only on the rational behavior of the players—it drastically simplifies the realities of the superpower arms race.

Prisoners' Dilemma omits two salient features of the superpower arms race that we believe need to be incorporated to produce a more realistic model, the focus of our attention in the remainder of this chapter. First, a player does not make a dichotomous choice between cooperation (disarming) and noncooperation (arming) but rather chooses a kind or level of action, or arms expenditures, that may be interpreted as being escalatory or deescalatory. Second, in response to an initial choice viewed as escalatory by his opponent, a player who was not viewed as escalatory at the start may subsequently choose a new level of expenditures that itself may be seen as escalatory or not.

In effect, players in the Deescalation Game described in the next section can choose initially to provoke or not provoke an opponent at any level; if provoked, a cooperative player can retaliate or not retaliate at any level. Thereby we incorporate into our model not only quantitative choices of any level of cooperativeness or noncooperativeness but also sequential choices that permit players to respond if provoked. The additional structure of quantitative and sequential choices in Prisoners' Dilemma not only better mirrors, in our view, real-world choices in the superpower arms race, but it also enables us to derive conditions under which it it rational for the players to be cooperative in the Deescalation Game and thereby escape the (2,2) trap.

2.3 THE DEESCALATION GAME

The Deescalation Game is defined by the following rules:

1 The players do not choose initially between C and $\bar{\text{C}}$, as in Prisoners' Dilemma, but instead choose (unspecified) actions that have associated with them a nonescalation probability (s for Row and t for Column) and a complementary escalation probability ($1-s$ for Row and $1-t$ for Column). With these probabilities, the actions will be interpreted as cooperative (C) and noncooperative ($\bar{\text{C}}$) moves, respectively.

2 Each player also chooses an (unspecified) retaliatory action that determines his conditional retaliation probability. If *both* players' initial choices are perceived as the same, the game ends at that position (i.e., CC or $\bar{\text{C}}\bar{\text{C}}$). But if one player's choice is perceived as C and the other's as $\bar{\text{C}}$, the first player's conditional nonretaliation probability (p for Column and q for Row) and complementary retaliation probability ($1-p$ for Column and $1-q$ for Row) come into play. With the retaliation probability, the conflict is escalated to the final outcome $\bar{\text{C}}\bar{\text{C}}$; otherwise it remains unchanged at $\text{C}\bar{\text{C}}$ or $\bar{\text{C}}\text{C}$.

3 The final outcome is one of the four outcomes of Prisoners' Dilemma—CC, $\text{C}\bar{\text{C}}$, $\bar{\text{C}}\text{C}$, or $\bar{\text{C}}\bar{\text{C}}$—determined as indicated in rule 2. The payoffs are the same as those of Prisoners' Dilemma, except that cardinal utilities replace ordinal rankings. Thus r_4 and c_4 signify the highest payoffs for Row and Column, respectively; r_1 and c_1 the lowest; etc.

It should be noted that the players choose both their escalation probabilities and retaliation probabilities at the start of the Deescalation Game, which is shown in figure 2.2. Note that besides the fact that the initial strategy choices of the two players are probabilities (with assumed underlying actions), rather than actions (C and $\bar{\text{C}}$) themselves, this payoff matrix differs from the figure 2.1 outcome matrix in having expected payoffs rather than (certain) payoffs as its off-diagonal entries. This is because we assume that if one player is perceived to escalate, the other player's (probabilistic) retaliation will be virtually instantaneous (as discussed in section 1.2). Hence, it is proper to include in the off-diagonal entries a combination of payoffs—reflecting both possible retaliation and possible nonretaliation —by means of an expected value.

We assume, of course, that $0 \le s, t, p, q \le 1$ because they represent probabilities. To simplify subsequent calculations, we normalize the payoffs of the players so that the best and worst payoffs are 1 and 0, respectively. Hence,

Column

		t	$1-t$
Row	s	(r_3,c_3)	$q(r_1,c_4)+(1-q)\,(r_2,c_2)$ $= ([1-q]r_2,q+[1-q]c_2)$
	$1-s$	$p(r_4,c_1)+(1-p)\,(r_2,c_2)$ $= (p+[1-p]r_2,[1-p]c_2)$	(r_2,c_2)

Key: $(r_i,c_j) =$ (payoff to Row, payoff to Column)
        $r_4,c_4 =$ best; $r_3,c_3 =$ next best; $r_2,c_2 =$ next worst; $r_1,c_1 =$ worst
        $s,t =$ probabilities of nonescalation
        $p,q =$ probabilities of nonretailiation
        Normalization: $0 = r_1<r_2<r_3<r_4 =1$; $0 = c_1<c_2<c_3<c_4 = 1$

Figure 2.2   Deescalation Game.

$0 = r_1 < r_2 < r_3 < r_4 = 1$
$0 = c_1 < c_2 < c_3 < c_4 = 1$

Because we assume that the escalation and retaliation probabilities are chosen simultaneously and independently by the players, the expected payoffs for Row and Column are simply the sums of the four payoffs (or expected payoffs) in the figure 2.2 matrix, each multiplied by its probability of occurrence:

$$E_R(s,q;t,p) = str_3+(1-s)t\,[p+(1-p)r_2]+s\,(1-t)\,(1-q)r_2$$
$$+(1-s)\,(1-t)r_2 \tag{2.1}$$
$$E_C(t,p;s,q) = stc_3+(1-s)t(1-p)c_2+s(1-t)\,[q+(1-q)c_2]$$
$$+(1-s)\,(1-t)c_2 \tag{2.2}$$

The introduction of escalation and retaliation probabilities into the expected payoff calculations requires some explanation and interpretation. Essentially we asume that every initial action that a player may take carries with it a probability of being interpreted as escalatory by his opponent and, if it is, a probability of drawing a response. This response, like the initial action that may escalate the conflict, is probabilistic in that it is not certain to constitute retaliation. Rather, both initial actions and subsequent responses have probabilities associated with their being viewed as escalatory and retaliatory, respectively, thereby leading to different outcomes in the game.

Thus, for example, the probability that Row will provoke Column by his choice is given by the escalation probability $1-s$. If Column is provoked, and providing that he did not also provoke Row initially (with escalation probability $1-t$), Column will respond with a subsequent action that further escalates the conflict to mutual noncooperation with retaliation probability $1-p$.

If neither player provoked the other (with probability st) or each provoked the other [with probability $(1-s)(1-t)$], then the retaliation probabilities never come into play, for we assume that there is (1) no need to retaliate for the choice of CC and (2) no possibility of retaliating for the choice of $\overline{C}\overline{C}$. Hence, the first and last terms of E_R and E_C given by (2.1) and (2.2) do not include retaliation probabilities.

The strategic problem that the players face is to choose both an initial level of action (with an associated escalation probability) and a subsequent level of response (with an associated retaliation probability). We assume, in interpreting the probabilities in the Deescalation Game, that the higher the level of (initial) escalation or (subsequent) retaliation, the greater the probability that these actions will be perceived as escalatory or retaliatory. Formally, then, we assume a linkage between the degree of escalation or retaliation and the probability that it will be interpreted as such by one's opponent.

When making their choices of initial and subsequent levels of action, and hence probabilities, before play of the game, we assume that the players know that their opponents will judge the level of these actions exactly as they do themselves. Consequently, each player's probability assessment of each level of action will coincide with his opponent's. These probabilities become common knowledge once the levels of action to which they correspond are selected in the Deescalation Game.

With respect to the retaliation probabilities, it should be noted that they are not assumed to be a function of the escalation probabilities. To be sure, the higher one player's escalation probability, the more likely his opponent's retaliation probability will come into play and hence the more likely retaliation will occur. But since the retaliation as well as the escalation probabilities are chosen at the start of the game, the former (for one player) is necessarily independent of the latter (for the other).[3]

It is fair to ask why retaliation is ever a problem in Prisoners' Dilemma; it would seem, on the contrary, always to be a rational response by a player once he perceives his opponent has escalated the conflict by choosing $\overline{C}$. In the case of Row, for example, if Column has escalated to (4,1), he (Row) does immediately better by moving the game to (2,2), from which neither player would have an incentive to depart.

On the other hand, as we will show in chapter 3, if the game is Chicken, threatening to retaliate to $\overline{C}\overline{C}$, the mutually worst (1,1) outcome, would appear irrational. We argue, however, that the players can and do solve this problem by precommitting themselves to carry out threats, despite the apparent irrationality of doing so in terms of the consequences for the threatener. We will analyze what operational form precommitments might take and, indeed, whether they are consistent with precepts of rational play that the rules of a game permit.

We assume that the same kind of precommitments to retaliate can be made in the case of the Deescalation Game. In this game, however, it is the *combination* of escalation and retaliation probabilities that may make initial escalation for, say, Row, from (r_3,c_3)—rather than subsequent retaliation by Column from (r_4,c_1)—irrational, should the players end up at (r_2,c_2) sufficiently often.

Without an adequate precommitment to retaliate on the part of Column, Row may think that he can escalate without serious or probable repercussions, although this subjects Column to his worst outcome. In our model, however, we endow Column with the ability to choose a retaliation probability great enough to assure Row that "too high" an escalation probability on his part will be irrational, for it would carry the game from (r_3,c_3) through (r_4,c_1) to (r_2,c_2) "too often," yielding Row a smaller expected payoff from escalating initially rather than not escalating.

Put another way, a precommitment by one player to retaliate with a probability above a particular level—to be specified later—renders (prior) escalation on the part of the other unprofitable. This is a precommitment that, unlike in Chicken, seems unproblematic for the player responding to escalation. The problem is for the player who chooses escalation initially when it may, after possible retaliation, lead to a worse outcome.

It is precisely a player's precommitment to retaliate, as we will show, that has a surprising and salutary consequence in the Deescalation Game under certain conditions. Henceforth we will assume that players can precommit themselves to strategies—escalatory or retaliatory—so that there is never any doubt on the part of an opponent that they will be implemented, though perhaps with some probability less than 1.

The quantitative questions we next address in our game-theoretic analysis are what combinations of escalation and retaliation prob- abilities (1) maximize the payoff a player can guarantee for himself whatever his opponent does, (2) lead to Nash equilibria, and (3) induce cooperative choices that allow the players to escape the trap of mutual noncooperation. We in fact show that there are escape routes that justify deescalation as a rational policy; accordingly, we label our extension and refinement of Prisoners' Dilemma the Deescalation Game.

2.4 RATIONAL PLAY IN THE DEESCALATION GAME

Consider the Deescalation Game from Row's vantage point. In Prisoners' Dilemma, by choosing his dominant strategy C, he can guarantee himself a payoff of at least r_2, whatever Column chooses. This guaranteed minimum is Row's *security level*. By comparison, because Row chooses probabilities of certain actions and reactions, rather than strategies themselves, in the Deescalation Game, it is by no means obvious how much he can guarantee for himself independent of Column's (probabilistic) choices.

In the appendix to this chapter, we show that in fact Row can guarantee himself the same value as in Prisoners' Dilemma, namely r_2. We do this by calculating, first, the value of Row's expected payoff, E_R, when Column, by his choice of t and p, makes it as small as possible. We then assume that Row, by his choice of s and q, seeks to maximize this minimum value of E_R. The resulting value of E_R is Row's *maximin* value, or security level, for it is the value that Row can assure himself of even if Column seeks to minimize E_R.

There are two ways that Row can guarantee himself *at least* his maximin value: by choosing any of his strategies with $s = 0$, q arbitrary, or with $q = 0$, s arbitrary. In the former case, Row escalates with certainty. If Column also escalates or retaliates with certainty. Row obtains r_2; otherwise Row obtains a higher expected payoff, because it includes r_4 with some positive probability when Column does not retaliate. In the latter case, Row never escalates but always retaliates. If Column escalates with certainty, Row ensures himself of r_2; otherwise he obtains a higher expected payoff when Column does not escalate, because it includes r_3 with some positive probability.

Only when Column always escalates ($t = 0$) does Row suffer his security level of r_2 when he chooses any of his maximin strategies. When $t > 0$, by contrast, Row always can do better than r_2. In this case, however, which maximin strategy serves him best depends on Column's choice of p, as shown in the appendix. Column's maximin strategies and security level are analogous because of the symmetry of the Deescalation Game.

Maximin strategies, especially in variable-sum games like the Deescalation Game, are conservative in the extreme, for they presume that one's opponent desires only to minimize one's own payoff even if it hurts him to do so. By contrast, in constant-sum games, maximin strategies (which turn out also to be *minimax strategies*—minimizing one's opponent's maximum payoff) are more defensible because hurting the opponent always helps oneself.

If perhaps overly conservative, however, each player's maximin strategy of escalating with certainty,

$$s = 0, q \text{ arbitrary}; \qquad t = 0, p \text{ arbitrary} \tag{2.3}$$

results in a Nash equilibrium, which we call the *escalation equilibrium*. This equilibrium, of course, corresponds to the unique Nash equilibrium of (r_2, c_2) in Prisoners' Dilemma. Since a player who always escalates forgoes any opportunity ever to retaliate in the Deescalation Game, the escalation equilibrium is independent of whatever retaliation probabilities (q and p) the players choose in this game, which is why we designate them as "arbitrary."

Auspiciously, the escalation equilibrium is not unique in the Deescalation Game. As shown in the appendix, there is a second Nash equilibrium,

$$s = 1, q \le \frac{c_3 - c_2}{1 - c_2}; \qquad t = 1, p \le \frac{r_3 - r_3}{1 - r_2} \tag{2.4}$$

which we call the *deescalation equilibrium*. It says that a player (say, Column) will never escalate ($t = 1$); but in response to escalation by Row, sometimes Column will not retaliate [with nonretaliation probability $p \le (r_3 - r_2)/(1 - r_2)$] and at other times he will [with retaliation probability $1 - p \ge (1 - r_3)/(1 - r_2)$]. More accurately, Column will choose actions in response to any prior (escalatory) actions by Row with a retaliation probability equal to or greater than the threshold value, $(1 - r_3)/(1 - r_2)$.

Why this threshold value? As shown in the appendix, this is the retaliation probability of Column that makes Row's expected payoff, E_R, independent of his choice of s. If Column's retaliation probability exceeds this threshold, however, Row would (irrationally) decrease E_R should he deviate from $s = 1$ (i.e., by choosing $s < 1$). Hence, given that Column's retaliation probability is above this threshold, Row maximizes E_R by choosing $s = 1$ and will *not* have an incentive to deviate. For analogous reasons, Column will not deviate from the deescalation equilibrium, rendering the resulting outcome stable. This outcome, of course, corresponds to the (r_3, c_3) compromise in Prisoners' Dilemma.

Perhaps the most significant feature of the Deescalation Game is that it makes the compromise outcome stable, even though this outcome is highly unstable in the underlying Prisoners' Dilemma game. This stability is due to the fact that the values of the two off-diagonal outcomes of Prisoners' Dilemma, which give Row rank 4 (best) at one outcome (lower left in figure 2.1) and Column rank 4 at the other (upper right in figure 2.1), are diminished in the Deescalation Game to expected values no more than r_3 and c_3 by the deescalation equilibrium strategies.

In effect, the probability of retaliation at this equilibrium dilutes the value of a win, r_4 or c_4, with the value of the much less desirable trap outcome, r_2 or c_2. Since the payoffs at the compromise outcome, r_3 and c_3, are unaltered in the Deescalation Game from those in Prisoners' Dilemma, they become, in relative terms, the most attractive when retaliation is sufficiently likely at the deescalation equilibrium. That is, when both sides are prepared to retaliate, nonescalation is each player's best strategy, and compromise the mutually best outcome.

In the appendix we conduct an exhaustive search for Nash equilibria in the Deescalation Game, showing that there are none other than the escalation equilibrium and the deescalation equilibrium. One effect, then, of retaliation probabilities above the threshold in this game is to transform the cooperative outcome from a next-best nonequilibrium (in the underlying Prisoners' Dilemma game) to a best equilibrium (in the Deescalation Game), without changing the payoffs to the players. The deescalation equilibrium, however, is not the product of dominant strategies in the Deescalation Game: One player's deescalation equilibrium strategy is best if the other player chooses his but definitely not best if the other player chooses certain other strategies.

At the same time that (r_3,c_3) is stabilized in the Deescalation Game, the stability of (r_2,c_2) is called into question, even though it corresponds to a Nash equilibrium. To see why, assume that the players begin at the outcome defined by

$$s = 0, q = q_0; \quad t = 0, p = p_0; \quad \text{or } (0,q_0;0,p_0) \tag{2.5}$$

where p_0 and q_0 are arbitrary. Since the players escalate with certainty, they receive payoffs (r_2,c_2) at this escalation equilibrium.

Now let Column change his strategy to $t = t_0$, $p = 0$, so the strategies become

$$(0,q_0;t_0,0) \tag{2.6}$$

where Column escalates with arbitrary probability $t_0 > 0$ and always retaliates ($p = 0$). The players still receive (r_2,c_2), but Column has changed his Nash equilibrium strategy [i.e., his probabilities in (2.5)] without cost to himself.

If Row next changes his Nash equilibrium strategy to never escalate but always retaliate, giving

$$(1,0;t_0,0) \tag{2.7}$$

his expected payoff will be

$$E_R(1,0;t_0,0) = t_0 r_3 + (1 - t_0)r_2$$

This is clearly better for him (since $t_0 > 0$) than r_2 that he receives at the deescalation equilibrium and at (2.6), so he would be motivated to switch from (2.6) to (2.7). In fact, switching from $s = 0$, $q = q_0$ to $s = 1$, $q = 0$ maximizes Row's expected payoff as long as Column plays $t = t_0$, $p = 0$.

But now, if $t_0 < 1$, Column can respond to the situation at (2.7) by changing his strategy also, to never escalate but always retaliate, giving

$$(1,0;1,0) \tag{2.8}$$

This raises E_C for him from

$$E_C(t_0,0;1,0) = t_0 c_3 + (1 - t_0)c_2$$

at (2.7) to

$$E_C(1,0;1,0) = c_3$$

at (2.8), which is the deescalation equilibrium with payoffs (r_3, c_3) for both players. Again, Column's move from (2.7) to (2.8) maximizes his return, assuming Row's strategy is fixed.

Thereby the players can move progressively along the path defined by

$$(2.5) \xrightarrow[\text{to Column}]{\text{Costless}} (2.6) \xrightarrow[\text{to Row}]{\text{Beneficial}} (2.7) \xrightarrow[\text{to Column}]{\text{Beneficial}} (2.8)$$

with only the first step that triggers the process not positively beneficial to the player (Column) who makes the initial move from the escalation equilibrium. But it is a costless change for Column,[4] so presumably he will make it if he anticipates that it will trigger the subsequent (beneficial) moves by Row and Column, respectively.

Indeed, the "trigger condition" can be relaxed to $t_0 > 0$, $p_0 < (r_3 - r_2)/(1 - r_2)$ in (2.6) in the sense that any such (t_0, p_0) chosen by Column would motivate Row to choose $s = 1$, $q = 0$ at (2.7). However, use of any p_0 satisfying $0 < p_0 < (r_3 - r_2)/(1 - r_2)$ would reduce (temporarily) Column's payoff to

$$E_C(t_0,p_0;0,q_0) = c_2 - t_0 p_0 c_2$$

As noted previously, $p_0 = 0$ is costless, so the $(2.5) \rightarrow (2.6) \rightarrow (2.7) \rightarrow (2.8)$ path is the most persuasive: No player would ever suffer any loss in departing from his Nash equilibrium strategies, making the need for irrevocable precommitments less. Obviously, the roles of Column and Row can be reversed to trace another path from the escalation to the deescalation equilibrium.

It is interesting to note in the Deescalation Game that it is the escalation equilibrium that exhibits some instability, for a costless perturbation by one player induces an immediate shift away from the escalation equilibrium toward the deescalation equilibrium. The perturbation triggering the shortest path to deescalation is $t = t_0 = 1$, $p = 0$. It is also noteworthy that this particular perturbation strategy—never escalate, always retaliate—bears a strong resemblance to the tit-for-tat strategy recommended by Axelrod for iterated Prisoners' Dilemma.[5]

We do not, however, assume repeated play of Prisoners' Dilemma but only an ability to retaliate for an initial untoward action of an opponent. Remarkably, this retaliatory ability turns out to be sufficient both to deter an opponent and to induce him to shift to the same deterrent strategy, from which he also will benefit. Once both players have adopted—and precommitted themselves to—this posture, then not only are their payoffs at the deescalation equilibrium better than at the Pareto-inferior escalation equilibrium but also the new equilibrium is highly stable: Both players would do immediately worse by deviating from $s = t = 1$ (never escalate) because of possible retaliation.

However, the players can afford to raise $p = q = 0$ (certain retaliation) up to the threshold values given by (2.4) for the deescalation equilibrium, thereby making retaliation less than certain, and still maintain stability. In other words, each player's retaliatory threat need only be probabilistic, and so can be taken to have as its *certain equivalent* a lower-level retaliatory action that would make each indifferent between choosing a lottery (over full-scale retaliation/non-retaliation) and a sure thing (the lower level of retaliation).

A less-than-certain or lower-level threat may be more credible, especially in the Deterrence Game based on Chicken that we analyze in chapter 3, wherein the threat may be escalation to nuclear war. By contrast, in arms races modeled by the Deescalation Game, and in Prisoners' Dilemma on which it is based, each player has an evident incentive to carry out a threat because he immediately benefits—by protecting himself from being overtaken in such a race—even if the resulting outcome is Pareto-inferior.

In fact, whether the underlying game is Chicken or Prisoners' Dilemma, the purpose of threatening retaliation is to deter an opponent from deviating from (r_3, c_3), regardless of whether it is costly to carry out the threat. Thus, the logic underlying threats that stabilize (r_3, c_3) in both games, which are special cases of the Conflict Game in chapter 1, is exactly the same. But beyond the use of retaliatory threats to render this outcome an equilibrium in the Deescalation Game, we believe even more hopeful is our finding that there is a costless, and in general beneficial, way for the players to escape the (r_2, c_2) trap and reach the (r_3, c_3) compromise outcome in this game.

2.5 CONCLUSIONS

Not only are arms races terribly costly but they may also increase the probability of war between two states under certain conditions.[6] When these states are the superpowers, and the costs are in the hundreds of billions of dollars—with nuclear holocaust a possible consequence of fighting that may erupt in an extreme crisis—then there is good reason to ponder how to deescalate the superpower arms race.

The arms race has persisted, we believe, because both sides see it as a Prisoners' Dilemma, with little hope of escaping the (2,2) trap. To be sure, the superpowers have been able to reach some arms control agreements. For the most part, however, the agreements hve been of a very limited nature, and even some of these are in trouble today because of mistrust and suspicions of cheating as well as new technological developments.[7]

The Deescalation Game, insofar as it reflects the quantitative choices about arms expenditures that each side makes—and the possible responses to the other side's perceived expenditures—give some basis for being sanguine. First, by stabilizing the compromise outcome (r_3, c_3), and second, by showing that there is a rational path from the trap (r_2, c_2) to the compromise (r_3, c_3), it suggests how the deescalation equilibrium might supplant the escalation equilibrium as its rational outcome. Essentially each side must precommit itself to respond to, but not initiate, escalation. Retaliation, while rational in Prisoners' Dilemma (as opposed to Chicken) once one side has escalated, nevertheless hurts both players at the resulting escalation equilibrium, at least in comparison to the deescalation equilibrium.

The fact that the players can extricate themselves from the escalation equilibrium by a series of rational moves and responses in the Deescalation Game is what makes this game a much more pleasant one to play than Prisoners' Dilemma. If it is also a more realistic model of Prisoners' Dilemma–type conflicts—such as the superpower arms race—then it suggests a solution, at least at a conceptual level, to the pathology of such conflicts when they have the quantitative, sequential character of the Deescalation Game.

We think that arms races, particularly the arms race between the superpowers, have this character. Of course, it will require a leader of imagination and courage to commit himself to deescalatory policies, though our model suggests that to be effective he must combine these with the threat of possible retaliation if the other side does not follow suit. Given such a carrot-and-stick combination, this posture does not require any great daring, because, at least in theory, it is costless. In reality, this may not be entirely so—for domestic political reasons, among others—but we believe our model goes a considerable way toward justifying a more conciliatory posture if the threat of retaliation is palpable.

Unfortunately, our model is not one that can be readily tested, because, at least in the superpower arms race, there has been no major deescalation. Moreover, this largely offensive arms race would surely be complicated, and probably aggravated, by a defensive arms race, whose strategic effects we investigate in our analysis of Star Wars in chapter 5.

Since conditions under which deescalation has actually occurred are rare, we think the model's value is more normative than explanatory. It suggests a path leaders might choose, and the kinds of threats necessary to induce one's adversary to assent to cooperation, that could lead to arms reductions. Whether they occur, we believe, may offer a test of the model's applicability to prospective rather than retrospective understanding.

APPENDIX

We present the details of our analysis of the Deescalation Game in this appendix. We begin by calculating the players' maximin strategies and values, and then we determine all Nash equilibria by an exhaustive search.

The rules of the Deescalation Game are given in the text, where the payoffs and strategic choices, and their interpretations, are made explicit. The game is depicted in figure 2.2. For convenience, the expected payoffs of Row (R) and Column (C) are repeated here:

$$
\begin{aligned}
E_R(s,q;t,p) &= str_3 + (1-s)t[p+(1-p)r_2] + s(1-t)(1-q)r_2 \\
&\quad + (1-s)(1-t)r_2 \qquad\qquad\qquad (1) \\
&= r_2 + st(r_3-r_2) + (1-s)tp(1-r_2) - s(1-t)qr_2 \\
E_C(t,p;s,q) &= stc_3 + (1-s)t(1-p)c_2 + s(1-t)[q+(1-q)c_2] \\
&\quad + (1-s)(1-t)c_2 \qquad\qquad\qquad (2) \\
&= c_2 + st(c_3-c_2) - (1-s)tpc_2 + s(1-t)q(1-c_2)
\end{aligned}
$$

To identify Row's maximin strategy, suppose first that s and q are fixed and notice from (1) that

$$
\frac{\partial E_R}{\partial t} = s(r_3-r_2) + (1-s)p(1-r_2) + sqr_2 \geqq 0
$$

with equality iff $s = 0$ and $p = 0$. Thus, if Row chooses $s > 0$,

$$\min_{t,p} E_R(s,q;t,p) = \min_p E_R(s,q;0,p) = \min_p \{r_2 - sqr_2\} = r_2(1-sq)$$

Also, since $t \geqq 0$, $p \geqq 0$, and $r_2 < 1$,

$$\min_{t,p} E_R(0,q;t,p) = \min_{t,p} \{r_2 + tp(1-r_2)\} = r_2$$

Therefore, for all permissible values of s and q,

$$\min_{t,p} E_R(s,q;t,p) = r_2(1-sq)$$

so that Row's maximin value is

$$\max_{s,q} \min_{t,p} E_R(s,q;t,p) = \max_{s,q} \{r_2(1-sq)\} = r_2$$

Furthermore, Row can achieve his maximin value r_2 by choosing any of his strategies with $s = 0$ and q arbitrary, or with $q = 0$ and s arbitrary. It is interesting to note that any of these maximin strategies yields to Row exactly his maximin value when $t = 0$ [see (1)]; if $t > 0$, Row may receive more. Specifically, if $t > 0$ and $p > (r_3-r_2)/(1-r_2)$, a maximin strategy of the form ($s = 0$, q arbitrary) gives Row his best payoff, whereas if $t > 0$ and $p < (r_3-r_2)/(1-r_2)$, Row's preferred maximin strategy is $s = 1$, $q = 0$. If $p = (r_3-r_2)/(1-r_2)$, Row would be indifferent among his maximin strategies because

$$E_R = r_2 + t(r_3-r_2) - s(1-t)qr_2$$

which yields the same value, $r_2 + t(r_3-r_2)$, in every case.

It follows from the symmetry of the Deescalation Game that Column's maximin value is c_2, and that all Column's strategies with $t = 0$ or $p = 0$ are maximin strategies, guaranteeing him a payoff of at least c_2. The properties of Column's maximin strategies are analogous to those of Row's, as discussed above.

We turn now to the search for Nash equilibria in the Deescalation Game. Our search is organized according to the values of s and t at the equilibrium.

Case 1: $t = 0$

If $t = 0$, then (1) becomes

$$E_R(s,q;0,p) = r_2(1-sq)$$

so that Row's best reply to $t = 0$ is either $s = 0$, q arbitrary, or $q = 0$, s arbitrary. By symmetry, $t = 0$, p arbitrary, is also a best reply for Column against $s = 0$. It is easy to verify directly that all strategy combinations

$$s = 0, q \text{ arbitrary;} \qquad t = 0, p \text{ arbitrary} \qquad (3)$$

are equilibria. We call (3) the *escalation equilibrium*, since it is characterized by both players' escalating with certainty. At the escalation equilibrium, the outcome of the Deescalation Game is always the trap outcome of the underlying Prisoners' Dilemma game, yielding the players (r_2, c_2).

We now show that (3) are the only equilibria consistent with case 1 by considering Column's response to Row's strategy choice $s > 0$, $q = 0$. By (2),

$$E_C(t, p; s, 0) = c_2 + st(c_3 - c_2) - (1 - s)tpc_2$$

If $0 < s < 1$, then

$$\max_{t,p} E_C(t, p; s, 0) = \max_t E_C(t, 0; s, 0)$$
$$= \max_t \{c_2 + st(c_3 - c_2)\} = c_2 + s(c_3 - c_2)$$

which occurs at $t = 1$, $p = 0$. If $s = 1$, this maximum is also $c_2 + s(c_3 - c_2)$, occurring at $t = 1$. Therefore, Column's best reply to $s > 0$, $q = 0$, includes $t = 1$, so that no strategy combination including $s > 0$, $q = 0$, and $t = 0$ (as assumed in case 1) is an equilibrium.

Case 2: $s = 0$

By an argument analogous to that for case 1, the only equilibrium consistent with $s = 0$ is the escalation equilibrium (3).

Case 3: $t = 1$

If $t = 1$, then (1) becomes

$$E_R(s, q; 1, p) = r_2 + s(r_3 - r_2) + (1 - s)p(1 - r_2)$$
$$= [r_2 + p(1 - r_2)] + s[r_3 - r_2 - p(1 - r_2)] \qquad (4)$$

From the final expression of (4) it follows that $s = 1$ is Row's best reply if

$$p \le \frac{r_3 - r_2}{1 - r_2}$$

Symmetry places an analogous condition on q in order that $t = 1$ be Column's best response to $s = 1$. It is easy to verify directly that

$$s = 1, \ q \le \frac{c_3 - c_2}{1 - c_2}; \qquad t = 1, \ p \le \frac{r_3 - r_2}{1 - r_2} \tag{5}$$

is an equilibrium, which we call the *deescalation equilibrium*. Observe that the deescalation equilibrium always results in the compromise outcome of the underlying Prisoners' Dilemma game, yielding the players (r_3, c_3).

To show that there are no equilibria other than (5) consistent with case 3, we note first that if $p \ge (r_3 - r_2)/(1 - r_2)$, (4) implies that $s = 0$ would be Row's best reply; but we have already proved (case 2) that there are no equilibria with $s = 0$ and $t = 1$. The only remaining possibility is the combination $0 < s < 1$, $t = 1$, and $p = (r_3 - r_2)/(1 - r_2)$. But now (2) gives

$$\frac{\partial E_C}{\partial p} = -(1 - s)tc_2 < 0$$

since $0 < s < 1$ and $t = 1$. Thus $p = 0$ at any equilibrium with $0 < s < 1$ and $t = 1$, contradicting the inference [from (4)] that $p = (r_3 - r_2)/(1 - r_2) > 0$.

Case 4: $s = 1$

As in case 3, the only equilibrium with $s = 1$ is the deescalation equilibrium (5).

Case 5: $0 < s < 1; \ 0 < t < 1$

In this case, it follows from (2) that

$$\frac{\partial E_C}{\partial p} = -(1 - s)tc_2 < 0$$

so that $p = 0$ is necessary at any equilibrium. Similarly, $q = 0$ is necessary also. But now (1) shows that

$$E_R(s, 0; t, 0) = r_2 + st(r_3 - r_2)$$

which, since $t > 0$, Row can maximize only at $s = 1$. Hence there are no equilibria consistent with case 5.

NOTES

This chapter is drawn from Steven J. Brams and D. Marc Kilgour, Rational deescalation, in *Evolution, Games, and Learning: Models for Adaptation in Machines and Nature*, ed. Doyne Farmer, Alan Lapedes, Norman Packard, and Burton Wendroff (Amsterdam: North-Holland, 1986), pp. 337–350, which is reprinted from *Physica D*, vol. 22. Reprinted here with permission from North-Holland Physics Publishing.

1 Technically, the players' payoffs must be in cardinal utilities rather than ordinal ranks in order for the game to be considered constant-sum or not. However, because Prisoners' Dilemma has a Pareto-inferior outcome, it follows that no assignment of utilities—necessarily less for both players at the Pareto-inferior outcome than at some Pareto-superior outcome—could possibly make it a constant-sum game.
2 Garthoff (1982, pp. 10–11). In a review of several different game-theoretic representations of the superpower arms race, Hardin (1983, p. 248) concluded that Prisoners' Dilemma reflects "the preference ordering of virtually all articulate policy makers and policy analysts in the United States and presumably also in the Soviet Union."
3 In chapters 6 and 7, in a model that posits different rules of play and new payoff functions, we assume that players do not choose retaliation probabilities that are constant for all levels of provocation but instead retaliation functions that depend on the level of provocation of one's opponent. In related work, Dacey (1987) analyzes within a decision-theoretic framework the use of probabilistic bribes, threats, and tit-for-tat combinations in a number of variable-sum games.
4 This is not the case in the Deterrence Game (discussed in chapter 3), wherein the player who triggers rational moves from a preemption equilibrium to the deterrence equilibrium will incur a temporary cost.
5 Axelrod (1984). Axelrod's analysis, however, is based on a very different game-theoretic model from ours. He found what when many computer programs representing strategies were matched against each other in tournament play of Prisoners' Dilemma, "tit-for-tat" did better than any other program. That is, if one starts out by cooperating but retaliates on the next round with noncooperation if the other player does not cooperate initially—if, in fact, one imitates the opponent's previous-round behavior in all subsequent play—then one does better on average than by never cooperating or by choosing most other strategies. (However, there is no strategy that is unequivocally best in tournament play; it depends on one's opponents' strategies.) Generally speaking, it pays to be "nice" (begin by cooperating), "provocable" (be ready to retaliate quickly if provoked), and "forgiving" (by returning to cooperating, after retaliating, as soon as the other player does). These conclusions, it should

be stressed, follow from a model that assumes repeated play against different (randomly chosen) opponents, which is hardly descriptive of superpower conflict that involves one continuing opponent. (In a personal communication to Brams, February 26, 1985, on this point, Axelrod suggested that, from the viewpoint of one superpower, the strategies of other players in a tournament might be conceptualized as representing that superpower's "probability distribution of the strategies of the other superpower"; but this conceptualization would require that each superpower have essentially the same view of his opponent—including an identical probability distribution over his strategies— which seems to us unrealistic.)

Possibilities for cooperation in iterated (but nontournament) play of Prisoners' Dilemma, with possible discounting of future payoffs, are investi- gated in, among other places, Taylor (1976) and Majeski (1984). In this work, as in Axelrod's, we find the dichotomous nature of choices (cooperation versus noncooperation) an unrealistic representation of arms race decisions, which we believe are less qualitative choices over an indefinite series of trials and more quantitative choices in single-play-with-the-possibility-of-retaliation sequences. Similarly, we think metagame theory, which posits qualitative choices contingent on the choices of others, is plagued both by being nonquantitative and by making heroic demands on the predictive capabilities of players about the choices of others. See Howard (1971); extensions and applications of metagame theory are given in Fraser and Hipel (1984) and Kilgour (1985). Finally, the ability to predict an opponent's choice, but not with certainty because of imperfect detection equipment, is incorporated in a two-stage Prisoners' Dilemma model in Brams, Davis, and Straffin, Jr. (1979a), which can also be found in Brams (1985b, ch. 3). In this work, it is shown that if each side has a sufficiently high probability of detecting the other side's (assumed) cooperative strategy choice in the first stage of the game (in the superpower arms race, for example, using satellite reconnaissance), a policy of conditional cooperation, based on tit-for-tat, leads to cooperation in the second stage. (The verification model of chapter 8 also assumes detection probabilities, but not in Prisoners' Dilemma.) Critiques of the Prisoners' Dilemma detection model are given in Dacey (1979) and Wagner (1983, p. 334). Responses to these critiques are given in Brams, Davis, and Straffin (1979b, 1984). For more on the question of tit-for-tat and its optimality, see chapter 6; in notes 2 and 3 of this chapter, we give a number of additional references to work inspired by Axelord, including applications of tit-for-tat to new games as well as variations that have been suggested in this kind of reciprocity policy.

6   There is by no means unanimity on this point. For example, Intriligator and Brito argue, on the basis of a dynamic model of a missile war from which they derive conditions of stable deterrence, that the chances of the outbreak of war have, paradoxically, been reduced because of the recent U.S.–Soviet quantita- tive arms race. See Intriligator and Brito (1984); for a critique and a response, see Mayer (1986) and Intriligator and Brito (1986). Recent reviews of the literature on the relationship between arms races and war, which include both theoretical and empirical assessments of this relationship, can be found in Morrow (1984) and Siverson (1986). In a more anecdotal vein, Kahn offers the following wry observation: "World War I broke out largely because of an arms

race, and World War II because of the *lack* of an arms race" (Kahn, 1984, p. 25; italics in original).

7 For discussions, see Brams (1985b, chs. 3 and 4); Smith (1985); and Nincic (1986).

3

Deterrence

3.1 INTRODUCTION

The policy of deterrence, at least to avert nuclear war between the superpowers, has been a controversial one. The main controversy arises from the threat of each side to visit destruction on the other in response to an initial attack. This threat would seem irrational if carrying it out would lead to a nuclear holocaust—the worst outcome for both sides. Instead, it would seem better for the side attacked to suffer some destruction rather than to retaliate in kind and, in the process of devastating the other side, seal its own doom in all-out nuclear exchange.

Yet the superpowers persist in adhering to *deterrence*, by which we mean a policy of threatening to retaliate for an attack by the other side in order to prevent such an attack in the first place. To be sure, nuclear doctrine for implementing deterrence has evolved over the years, with such appellations as "massive retaliation," "flexible response," "mutual assured destruction" (MAD), and "counterforce" giving some flavor of the changes in strategic thinking in the United States.

All such doctrines, however, entail some kind of response to a Soviet nuclear attack. They are operationalized in terms of preselected targets to be hit, depending on the perceived nature and magnitude of the attack (Ball and Richelson, 1986; Martel and Savage, 1986). Thus, whether U.S. strategic policy at any time stresses a retaliatory attack on cities and industrial centers (countervalue) or on weapons systems and armed forces (counterforce), the certainty of a response of some kind to an attack is not the issue. The issue is, rather, what kind of threatened response, or *second strike* in the parlance of deterrence theory, is most efficacious in deterring an initial attack, or *first strike*.

This is the issue we address in this chapter, though not in the usual way. Instead of trying to evaluate the relative merits of concrete nuclear retaliatory doctrines, we define these doctrines somewhat more abstractly in terms of "probabilistic threats." As in chapter 2, by letting threats vary along a single continuous dimension from certain retaliation to no retaliation, we can compare different levels of threats in terms of the expected payoffs that they yield in a game. Additionally, by introducing probabilities of a first strike (or preemption) by both sides—analogous to the probabilities of escalation in chapter 2—we can analyze the relationship between preemption and retaliation probabilities on the one hand, and game outcomes on the other.

As before, these probabilities can be interpreted as *levels* of preemption and retaliation that may fall short of full-fledged first and second strikes. The first question we seek to answer is what levels of preemption and retaliation render certain outcomes stable, or Nash equilibria.

In the game we use to model deterrence, which is derived from the game of Chicken but is not Chicken itself, we identify four stable outcomes, or equilibria, three of which correspond to those in Chicken. The new equilibrium, which emerges when we incorporate the possibility of (probabilistic) preemption and retaliation into Chicken, we call the *deterrence equilibrium.* It corresponds to the cooperative outcome in Chicken (never preempt), which by itself is unstable; in the new game, which we call the *Deterrence Game*, this outcome is rendered stable by the threat of retaliation above a calculable threshold, which makes preemption irrational.

But a threshold alone does not specify what level of threat (above this threshold) is optimal. Accordingly, we suggest a theoretical calculation of "robust threats" that make retaliatory threats as invulnerable as possible to misperceptions or miscalculations by the players. We also indicate how precommitments to carry out these threats are in fact made credible, at least on a probabilistic basis, by the superpowers.

We think the deterrence equilibrium, supported by robust threats, is superior to any other equilibrium in the game we postulate as a model of deterrence. To be sure, this equilibrium is "imperfect" in the technical sense that it is irrational to carry out one's threats; however, because it renders preemption irrational, even when one thinks one's opponent might preempt, it is hard to see why retaliation would ever be necessary, at least in theory. This theoretical rationale for a particular kind of deterrence, coupling a no-first-use policy with robust threats, appears to us the best one can do in a world that seems to make superpower confrontations unavoidable.

The challenge facing the policy maker is to prevent such confrontations from escalating into nuclear war, which we have more to say about in chapter 7. As deleterious as threats are to the development of trust and good will, we conclude here that they are inescapable for deterrence to be effective. It is far less clear whether the threats that the superpowers hurl at each other today, and their concomitant actions to indicate that the threats are not empty, are at an optimal level. However, we defer a full-fledged discussion of the optimal level of threats as a function of the level of provocation until chapter 6.

3.2 DETERRENCE AND CHICKEN

There is a large literature on deterrence, but little of it is explicitly game-theoretic. That which is, or is pertinent to game-theoretic formulations, is discussed by both Brams and Zagare from a theoretical as well as empirical perspective, so we will not review it here.[1] Suffice it to say that we believe not only that game theory provides a framework uniquely suited to capturing the interdependent strategic calculations of players but also that it can be adapted to modeling the threats necessary to deter an opponent from taking aggressive action against oneself.

To incorporate threats into the structure of a game, we assume that players can precommit themselves to carrying out their threats with a given probability. Exactly how they do so will be considered later, but for now we assume, as in chapter 2, that precommitments are allowed by the rules of the game.

Because a game is defined by the rules that describe it, there is no problem in permitting precommitments as long as they are not inconsistent with other rules. In fact, as we will show, the major issue precommitments raise is whether it is rational to hold to them in the play of a game. We will discuss this issue after deriving the equilibria of the Deterrence Game and analyzing their properties.

The Deterrence Game is based on the two-person game of Chicken. In Chicken, as in Prisoners' Dilemma, each player can choose between two strategies: cooperate (C) and not cooperate ($\bar{C}$), which in the context of deterrence may be thought of as "not attack" and "attack," respectively. These strategies lead to the four possible outcomes shown in figure 3.1:

1 Both players cooperate (CC): next-best outcome for both players— (3,3)
2 One player cooperates and the other does not (C$\bar{C}$ and $\bar{C}$C): best outcome for the player who does not cooperate and next-worst outcome for player who does—(2,4) and (4,2)
3 Both players do not cooperate ($\bar{C}\bar{C}$): worst outcome for both players—(1,1)

Outcomes (2,4) and (4,2) in Chicken are Nash equilibria. For example, from (2,4) Row would do worse if he moved to (1,1), and Column would do worse if he moved to (3,3). By contrast, from (3,3),

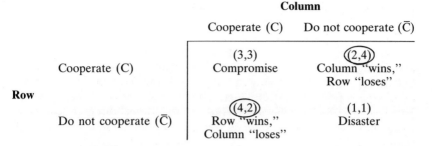

Figure 3.1 shows the outcome matrix of Chicken, with Row choosing between Cooperate (C) and Do not cooperate (C̄), and Column choosing between Cooperate (C) and Do not cooperate (C̄). The outcomes are:

- (C, C): (3,3) Compromise
- (C, C̄): (2,4) Column "wins," Row "loses"
- (C̄, C): (4,2) Row "wins," Column "loses"
- (C̄, C̄): (1,1) Disaster

Key: (x,y) = (rank of Row, rank of Column)
4 = best; 3 = next best; 2 = next worst; 1 = worst
Circled outcomes are Nash equilibria

Figure 3.1 Outcome matrix of Chicken.

Row would do better if he moved to (4,2), and Column would do better if he moved to (2,4), so (3,3) is unequivocally not a Nash equilibrium.

There is a third Nash equilibrium in Chicken, but it is in mixed strategies rather than pure strategies (see section 1.2). Because the calculation of equilibria involving mixed strategies requires that payoffs be given in cardinal utilities, not just ordinal ranks, we postpone discussion of these strategies and the third equilibrium until the development of the Deterrence Game in section 3.3.

The shorthand verbal descriptions given for each outcome in figure 3.1 suggest the vexing problem that the players confront in choosing between C and C̄: By choosing C̄, each can "win" but risks disaster; by choosing C, each can benefit from compromise but can also "lose." Each of the pure-strategy Nash equilibria shown in figure 3.1 favors one player over the other, but the stability of these equilibria as such says nothing about which of the two, if either, will be chosen.

Other concepts of equilibrium distinguish (3,3) as the unique stable outcome, but the rules of play that render compromise stable presume that the players (1) act nonmyopically or farsightedly and (2) cannot threaten each other (Brams and Wittman, 1981; Kilgour, 1984; Zagare, 1984). If threats are possible in repeated play of Chicken under still different rules, the stability of (3,3) is undermined (Brams and Hessel, 1984).

The effect that threats may have in Chicken is not hard to grasp. If one player (say, Row) threatens the other player (Column) with the choice of C̄, and this threat is regarded as credible, Column's best response is C, leading to (4,2).

Clearly, the player with the credible threat—if there is one—can force the other player to back down in order to avoid (1,1). Although

Row would "win" in this case by getting his best outcome, Column would not "lose" in the usual sense by getting his worst outcome but would instead get his next worst.

That is why we have put "win" and "lose" in quotation marks here and in figure 3.1. As in other variable-sum games (e.g., Prisoners' Dilemma), *both* players may do better at some outcomes [e.g., (3,3)] than others [e.g., (1,1)]. In fact, (1,1) is the unique Pareto-inferior outcome in Chicken; because none of the other three outcomes is better for both players than either of the other two, they are all Pareto-superior.

We have shown that Chicken is vulnerable to the use of *threats*, by which we mean a *precommitment* (before the play of the game) to the choice of a strategy by one player in order to force the other player to choose a strategy, and hence an outcome, favorable to the threatener. For a threat to be *effective* (i.e., force the threatened player to choose the strategy the threatener prefers), it must be *credible*: The threatened player must believe that the threatener will in fact carry out his threat.

Thus, for example, if Column did not believe that Row would actually choose $\bar{C}$ in Chicken (e.g., because he himself also threatened to choose $\bar{C}$), Column presumably would choose $\bar{C}$ in the belief that Row would back down and choose C, leading to Column's best outcome of (2,4). Of course, if Column's belief were mistaken, the outcome for both players would be disastrous. In the Deterrence Game, we will explore how mutual threats in Chicken may induce compromise rather than push the players toward the precipice.

Chicken is not the only game vulnerable to threats. There are 78 distinct strict ordinal 2×2 games in which two players, each with two strategies, can strictly rank the four outcomes from best to worst. In 46 of them, one or both players has "threat power" of either a "compellent" or "deterrent" kind.[2] Chicken, however, is the only one of the 78 games that satisfies the following four conditions:

1 *Symmetry*. The players rank the outcomes along the main diagonal (CC and $\bar{C}\bar{C}$) the same; their rankings of the off-diagonal outcomes (C$\bar{C}$ and $\bar{C}$C) are mirror images of each other.
2 *Cooperation is preferable to noncooperation*. Both players prefer CC to $\bar{C}\bar{C}$.
3 *Unilateral noncooperation helps the noncooperator and hurts the cooperator*. Row prefers $\bar{C}$C to CC to C$\bar{C}$, and Column prefers C$\bar{C}$ to CC to $\bar{C}$C.
4 *Retaliation for noncooperation is irrational*. If one player does not cooperate (i.e., the initial outcome is $\bar{C}$C or C$\bar{C}$), retaliation by the other player (to $\bar{C}\bar{C}$) is worse for the retaliator as well as for the player he retaliates against.

It is evident that all except conditon 1, which we assume in order to pose the same strategic dilemma for each player,[3] conspire to make Chicken a harrowing game to play. Cooperation is at the same time desirable (condition 2) and undesirable (condition 3). But the crux of the dilemma is that if one player is intransigent (i.e., noncooperative), the other player has good reason not to be intransigent (condition 4).

If condition 4 does not obtain, but instead $\overline{C}C$ is better than CC and $C\overline{C}$ for the cooperative player, then the resulting game is Prisoners' Dilemma, wherein the worst (1) and next-worst (2) outcomes of Chicken are interchanged for each player. As is apparent, Chicken presents its players with a very different kind of strategic problem from Prisoners' Dilemma, although, as we shall show, stabilizing the cooperative outcomes in both the Deescalation and Deterrence Games involves the same kinds of probabilistic threats.

The heart of the problem with deterrence, especially of the nuclear kind, is the apparent irrationality of retaliating against a first strike by an opponent.[4] What sort of threats (if any) are credible, and will deter a first strike, so as not to put one in the unenviable position of having to decide whether to retaliate and court mutual annihilation? When is a policy of deterrence involving mutual threats of retaliation stable? How can players make their precommitments to retaliate compelling? We explore these and other questions in our analysis of the Deterrence Game, which requires that the players choose, as in the Deescalation Game, both levels of preemption (escalation in the Deescalation Game) and levels of retaliation at the beginning of play.

3.3 THE DETERRENCE GAME

The Deterrence Game, shown in figure 3.2, is defined by exactly the same rules as those given for the Deescalation Game in section 2.3, except that "Chicken" is substituted for "Prisoners' Dilemma," and "preemption" for "escalation," in the game descriptions. The expected payoffs for Row and Column (after normalization), analogous to those given for the Deescalation Game in section 2.3, are as follows:

$$E_R(s,q;t,p) = str_3 + (1-s)tp + s(1-t)qr_2 \qquad (3.1)$$
$$E_C(t,p;s,q) = stc_3 + s(1-t)q + (1-s)tpc_2 \qquad (3.2)$$

In the appendix we show that there are effectively four Nash equilibria in the Deterrence Game, and that they can be grouped into three classes:

I *Deterrence equilibrium*:

$$s = 1, q \le c_3; \qquad t = 1, p \le r_3$$

Column

	t	$1-t$
s	(r_3,c_3)	$q(r_2,c_4)+(1-q)\,(r_1,c_1)$ $= (qr_2,q)$
$1-s$	$p(r_4,c_2)+(1-p)\,(r_1,c_1)$ $= (p,pc_2)$	$(r_1,c_1) = (0,0)$

Row

Key: (r_i,c_j) = (payoff to Row, payoff to Column)
r_4,c_4 = best; r_3,c_3 = next best; r_2,c_2 = next worst; r_1,c_1 = worst
s,t = probabilities of nonpreemption
p,q = probabilities of nonretaliation
Normalization: $0 = r_1<r_2<r_3<r_4 = 1$; $0 = c_1<c_2<c_3<c_4 = 1$

Figure 3.2 Deterrence Game.

This equilibrium is one in which the players never preempt ($s = t = 1$), but row retaliates with probability $1-p > r_3$ and Column retaliates with probability $1-q > c_3$. Essentially, these inequalities ensure that a player's expected payoff as the sole preemptor—p for Row and q for Column, as shown in the off-diagonal entries in figure 3.2—is not greater than what is obtained from the cooperative outcome of the underlying Chicken game, with payoffs (r_3,c_3).

II *Preemption equilibria*:

(1) $s = 1$, $q = 1$; $t = 0$, p arbitrary
(2) $s = 0$, q arbitrary; $t = 1$, $p = 1$

The first equilibrium is certain preemption by Column and no retaliation by Row; because Row is deterred by Column's initiative, Column's retaliation probability is arbitrary because it never comes into play. The second equilibrium is analogous, with the roles of Column and Row switched. At these equilibria, the outcomes of the Deterrence Game are the outcomes of the underlying Chicken game associated with wins for Column and Row (discussed in section 3.2), with payoffs $(r_2,1)$ and $(1,c_2)$, respectively.

III *Naive equilibrium*:

$$s = \frac{c_2}{1-c_3+c_2}, q = 1; \qquad t = \frac{r_2}{1-r_3+r_2}, p = 1$$

At this equilibrium, each player preempts with some nonzero probability (which depends on the other player's payoffs and is always less than 1) but never retaliates. Each of these preemption probabilities in fact (see appendix) makes the opponent indifferent as to his level of preemption; in other words, a player's expected payoff depends only on his opponent's, and not his own, level of preemption. Because retaliation would only degrade these expected payoffs, it is suboptimal. As shown in the appendix, however, the naive equilibrium is Pareto-inferior to the deterrence equilibrium, which is the reason for our nomenclature. It corresponds to the mixed-strategy equilibrium of the underlying Chicken game (discussed but not given in section 3.2), which is similarly deficient as well as difficult to interpret as a one-shot choice of rational players in this game.

Of the four Nash equilibria, only the deterrence equilibrium in class I depends on the possibility of retaliation—specifically, precommitted threats to respond (at least probabilistically) to a provocation when it is viewed as equivalent to the choice of $\bar{C}$. Such threats distinguish the Deterrence Game from the underlying game of Chicken, in which retaliation against the choice of $\bar{C}$ is not permitted.

Note that the two preemption equilibria in class II, and the one naive equilibrium in class III, occur only when retaliatory threats are never used ($p = 1$ or $q = 1$ or both). These equilibria correspond precisely to the three Nash equilibria in Chicken and so introduce no new element into the analysis of deterrence beyond what was earlier provided by Chicken. However, when a threat structure is added to Chicken to give the Deterrence Game, a qualitatively different equilibrium (the deterrence equilibrium) emerges in the latter game that demonstrates how threats—uncomfortable as they may be to live with—can work to the advantage of both players to stabilize the Pareto-superior cooperative outcome (r_3, c_3), which is unstable in Chicken without the possibility of retaliation.

Not only is this outcome a Nash equilibrium, but it is also a dominant-strategy Nash equilibrium when precommitted retaliation probabilities are fixed: Comparing the choice of C (with probability, or at level, s) with the choice of $\bar{C}$ (with probability, or at level, $1-s$) in figure 3.2, $r_3 \geq p$ if Column is viewed as not preempting (with probability t); and $qr_2 \geq 0$ if Column is viewed as preempting (with probability $1-t$). Thus, for Row, an initial cooperative choice is at least as good, and sometimes better, and similarly for Column. This dominance would seem a strong argument for choosing the deterrence equilibrium over the preemption equilibria.

3.4 RATIONAL PLAY IN THE DETERRENCE GAME

Because the deterrence equilibrium depends fundamentally on conditional threats of retaliation that would be costly for the threatener to implement, it is not perfect in the sense of Selten (1975).[5] Specifically, the threat of retaliation in the Deterrence Game makes the deterrence equilibrium *imperfect*: Having to carry out this threat if deterrence fails hurts the threatener, even though the threat itself is what is supposed to prevent deterrence from failing in the first place (i.e., make it a Nash equilibrium).

Despite this shortcoming, the deterrence equilibrium possesses a *dynamic stability* property that, once the equilibrium forms, should contribute to its persistence in repeated play. This property says that, given the players are at the deterrence equilibrium, if one player (say, Row) for any reason suspects that the other player (Column) may contemplate preemption, thereby rendering $t < 1$, he (Row) can still do no better than continue to choose $s = 1$.

In other words, even if Row thinks he might be preempted, he should nonetheless continue to refuse to preempt in order to keep his expected payoff at its maximum. This obviates the problem that Schelling called "the reciprocal fear surprise attack" that leads inexorably to preemption.[6]

We prove this dynamic stability property of the deterrence equilibrium in the appendix; it establishes, in effect, that any perceived departures of s or t from 1 will not initiate an escalatory process whereby the players are motivated to move closer and closer toward certain preemption. The fact that the deterrence equilibrium is impervious to perturbations in s or t means that the players, instead of being induced to move up the escalation ladder, will have an incentive to move down should one player deviate from $s = t = 1$.

The restoration of the deterrence equilibrium depends on probabilistic threats of retaliation that satisfy

$$0 < q < c_3, \qquad 0 < p < r_3 \tag{3.3}$$

But note that if deterrence should fail for any reason, it is irrational to retaliate, even on a probabilistic basis, because retaliation leads to a worse outcome for the threatener, having to carry out his threat, as well as for the player who preempted and thereby provoked retaliation.

The apparent irrationality of retaliating in the Deterrence Game is precisely what makes the deterrence equilibrium imperfect. Despite its imperfectness, it, like the preemption equilibria, has the essential equilibrium property of "mutually fulfilled expectations": If either player anticipates that his opponent will choose an equilibrium strategy associated with one of the four equilibria (including the two preemption equilibria), he never can do better than choose his *corresponding* equilibrium strategy (associated with this same equilibrium) to maximize his payoff.[7]

The problem of imperfectness crops up again when we try to chart a path from one of the two preemption equilibria to the deterrence equilibrium. The player who is preempted must risk moving from his next-worst to his worst outcome in order to induce the premptor to switch from preemption to nonpreemption. Accordingly, in trying to trace a path from preemption to deterrence analogous to the one we indicated from escalation to deescalation in the Deescalation Game (section 2.4), our analysis will focus on the least costly route.

To shift the equilibrium outcome of the Deterrence Game from preemption by one player to deterrence by both, we ask how the cycle of expectation-fulfillment can be broken through unilateral action. We are interested particularly in what paths, or trajectories, from a preemption equilibrium to the deterrence equilibrium can be induced through the use of threats.

It is worth noting that players would never have the opposite incentive: to move from the deterrence equilibrium to a preemption equilibrium. For once at the deterrence equilibrium, it is not rational for either player ever to preempt (by departing from $s = 1$ or $t = 1$), because the preemptor's resulting expected payoff would be lower, assuming that the nonretaliation probabilities are fixed. Moreover, even if one player departed from his strategy of nonpreemption, the other player still would not be induced to move because of this equilibrium's dynamic stability property. True, a player (say, Row, by choosing $q > c_3$) can make it rational for his opponent (Column) to preempt him, but destabilizing the deterrence equilibrium to suffer a worse outcome is hardly rational.

On the other hand, it may be rational for Row, once at the preemption equilibrium where he suffers his next-worst payoff (r_2), to threaten retaliation [to $(0,0)$] unless Column pledges never to preempt $(t = 1)$. How to induce Column to make this pledge and backtrack from preemption, where he obtains a payoff of 1, to nonpreemption, where his payoff is $c_3 < 1$, is the question we analyze next.

Before considering this problem in the Deterrence Game, however, it is useful to consider the question in a more general context. Suppose a game is at some equilibrium A, and one player can credibly announce that he is changing his strategy so that it is no longer consistent with A but is instead consistent with another equilibrium, B. Assuming that his announcement is believed (in practice, this may require that the new strategy actually be chosen in one or more plays), one would expect that the other player will eventually adjust his strategy so as to achieve equilibrium B.

Naturally, a player would not be interested in making such an announcement unless he received a greater payoff at B than at A. The problem, however, is that the initiating player may incur a cost, albeit temporary, after he chooses his new strategy while his opponent continues with his old strategy. Presumably, the price to the initiating player of this change of equilibria will depend on both the magnitude of the temporary cost (in any single play) and the length of time (number of plays) it takes the responding player to switch to the new equilibrium.

We assume that the initiating player seeks to find a move away from equilibrium A that sets off a "chain reaction" of rational moves that eventually results in equilibrium B. Subsequent moves are rational if they serve, in turn, to maximize the mover's payoff. Although in general it will be (temporarily) costly for the initiating player to move from an equilibrium, games like the Deescalation Game demonstrate, as we indicated in section 2.4, that there may be no temporary costs at all for the initiator.

Consider now the possible trajectories in the Deterrence Game from the preemption equilibrium in which Column preempts Row (PE_C) to the deterrence equilibrium (DE). (The analysis for paths from PE_R, in which Row preempts Column, to DE is analogous because of the symmetry of the Deterrence Game, so it need not be considered separately.) A move from PE_C to DE would yield Row a net increase in payoff of $r_3 - r_2$ and Column a net decrease of $1 - c_3$, so it is obviously Row who would have the incentive to initiate this change in equilibria.

Suppose the initial strategy choices of the players are

$$s = 1, q = 1; \qquad t = 0, p = p_0; \qquad \text{or } (1,1;0,p_0) \tag{3.4}$$

where p_0 is arbitrary. This means that Row never preempts but Column always does; because Row never retaliates after being preempted, these strategies indeed represent PE_C.[8]

Next suppose that Row changes his nonretaliation probability to $q = q_0 < 1$. The game is now at

$$(1,q_0;0,p_0) \tag{3.5}$$

with payoffs $q_0 r_2$ to Row and q_0 to Column. Thus, Row incurs a temporary cost of $r_2 - q_0 r_2 = (1-q_0)r_2$ in making the change from (3.4) to (3.5).

Now Column's expected payoff by (3.2) is

$$E_C(t,p;1,q_0) = tc_3 + (1-t)q_0 = q_0 + t(c_3 - q_0)$$

It is easy to see that Column maximizes E_C by choosing $t = 1$, provided that $q_0 < c_3$.

Assume $q_0 < c_3$. Then Column is motivated to choose $t = 1$, resulting in

$$(1,q_0;1,p_0). \tag{3.6}$$

But this is a deterrence equilibrium, given that Column takes the additional step of changing p_0, if necessary, so that $p_0 \leq r_3$ in (3.6). (This step offers no immediate benefit to Column, but it does deter possible future defections by Row.)

So far we have shown that Row can initiate a trajectory from the preemption equilibrium where he is preempted to the deterrence equilibrium by changing his nonretaliation probability from $q = 1$ to $q = q_0 < c_3$. Because his temporary cost in making this change is $(1-q_0)r_2$, Row would have an incentive to choose a value of q_0 as large as possible (subject to $q_0 < c_3$). If $q_0 = c_3 - \varepsilon$, where ε is a small positive number, then Row's temporary cost is minimized at $(1-c_3)r_2 + \varepsilon r_2$.

We show in the appendix that the choice of $s = 1$, $q = c_3 - \varepsilon$ is the most cost-effective way for Row to trigger a trajectory from PE_C to DE. That is, Row can set in motion a chain of events leading from PE_C to DE most cheaply by lowering his nonretaliation probability from $q = 1$ to $q = c_3 - \varepsilon$.

The threat to retaliate with a probability just sufficient to make it more costly for Column to continue to preempt (not cooperate) than not to preempt (cooperate) is risky, of course, because it could lead to Row's worst outcome if Column does not accede to the threat. But Column faces the same risk as Row, since the worst outcome for one player is the same as for the other in the Deterrence Game. Hence, it is advantageous for Column also to accept his next-best payoff (c_3) rather than an expected payoff—reflecting the threat—that mixes his best (1) and worst (0) payoffs and is less desirable than cooperation [because $1p + 0(1-p) = p < c_3$].

In short, the preempted player must take a chance to topple a preemption equilibrium and replace it with the deterrence equilibrium. At least, however, there is a rational response on the part of his opponent that he can trigger, and at a calculable minimum cost.

Because the rational path from a preemption equilibrium to the deterrence equilibrium that the preempted player can initiate entails certain risks, it is fair to ask how threats of retaliation can be strengthened so as to forestall preemption in the first place. We will suggest two ways, one theoretical and one practical.

In theory, all threats that satisfy inequalities (3.3), given that $s = t = 1$, define a deterrence equilibrium. But in the intervals defined by (3.3), which values of p and q should be used? Brams proposed, as most insensitive to misperceptions or miscalculations by the players, *robust threats*,

$$q = \frac{c_3 - c_2 r_3}{1 - c_2 r_2}, \qquad p = \frac{r_3 - r_3 c_3}{1 - c_2 r_2}$$

which are easily shown to satisfy (3.3).[9] Such threats, when carried out, are equally damaging to the preemptor and equally costly to the retaliator, whichever strategy (preempt or not) a player perceives his opponent as choosing. This property makes each player's preemption decision independent of his reading of his opponent's choice—the damage or cost will be the same whatever he chooses—and should serve to enhance the stability of the deterrence equilibrium.

A by-product of robust threats is that they render nonpreemption (strategies *s* and *t* in the figure 3.2 Deterrence Game) *strictly dominant*—better for each player whatever his opponent does—and hence unconditionally best. (*Non*strict dominance means at least as good for every strategy choice of an opponent, and strictly better for some choice.) This, of course, is not true of the C strategies in Chicken (figure 3.1), which are *undominated*—sometimes best (when the opponent chooses C̄) and sometimes not (when he chooses C).

In practice, the deterrence equilibrium depends on the credibility of threats satisfying (3.3). But how does a player persuade his opponent that he will retaliate if attacked, even though retaliation would be irrational at the time it is undertaken?

In the case of the superpowers, both the United States and the Soviet Union have instituted detailed procedures for responding to a nuclear attack that are designed to ensure, insofar as possible, that retaliation will occur, even if command, control, communication, and intelligence (C^3I) capabilities are damaged by the attack.[10] However, although each side promises that a first strike will inevitably be met by a second strike, there is significant uncertainty about each side's likely response

because of a number of operational factors, including problems related to identifying the attacker, identifying the magnitude of the attack, failures of weapons being used for the first time on a massive scale, problems of communication and control, lack of resolve, and the like. In light of these difficulties, both sides have, not surprisingly, resorted less to making probabilistic threats and more to employing their certain equivalents—usually controlled steps up the escalation ladder.

These . . . may be thought of as probabilistic threats insofar as they give an opponent a better idea of how close each side is moving toward full-scale retaliation—that is, they indicate more palpably the probability that the opponent will carry out a threat and what its expected damage will be. So far, fortunately, these probabilistic threats have been sufficient to persuade the two sides to back off, beyond a certain point, from continued escalation.[11]

We conclude that (1) the deterministic threats proclaimed by the superpowers today are, in truth, probabilistic (as we have modeled them), and (2) they have in fact deterred nuclear war. Moreover, there seems little doubt that both sides have precommitted themselves to retaliating, even if the resulting doomsday machines have built-in certainties because of possible technical failures in C^3I—some of which may be irremediable—and other factors of a more human kind (e.g., lack of will to order a second strike). It therefore seems appropriate to characterize the mechanism underlying nuclear deterrence today as a "probabilistic doomsday machine," for which we have proposed the acronym PDM (Brams and Kilgour, 1986b; 1987e).

3.5 CONCLUSIONS

Deterrence means threatening to retaliate against an attack in order to prevent it from occurring in the first place. It is widely held that only through continuing mutual deterrence has a nuclear confrontation of the superpowers been avoided. Yet the central problem with a policy of deterrence is that the threat of retaliation may not be credible if retaliation leads to a worse outcome—perhaps a nuclear holocaust—than a state would suffer from absorbing a limited first strike and not retaliating.

We analyzed the optimality of mutual deterrence by means of a Deterrence Game, in which each player chooses a probability (or level) of preemption, and of retaliation if preempted. The Nash equilibria, or stable outcomes, in this game duplicate those in the game of Chicken, on which it is based, except for a deterrence equilibrium, at which the players never preempt but are always prepard to retaliate against preemption with a probability above a calculable threshold. (The latter equilibrium in fact corresponds to the cooperative equilibrium of the Conflict Game—of which the Deterrence Game is a special case—studied in section 1.2.)

The deterrence equilibrium is Pareto-superior, dynamically stable, and, when supported by robust threats, as invulnerable as possible to misperceptions or miscalculations by the players. Furthermore, we showed that there is a rational path from each preemption equilibrium to the deterrence equilibrium, although it is not without potential cost to the initiator. For the preemptor to ignore a credible threat of the preempted player, however, is irrational, although in reality, of course, it may occur.

How do these results accord with the strategic doctrine of MAD? First, MAD is not only an acronym for "mutual assured destruction" but also for "mutual assured deterrence." In its former incarnation, MAD is more of an epithet than a statement of policy, except insofar as it implies that to save the world each side must be willing to destroy it.

Second, our deterrence equilibrium suggests that this is only partially true: There is not, and need not be, "assured destruction," but only a probabilistic threat of it to induce "assured deterrence." If the threat of retaliation is sufficiently great, and perceived to be credible, neither side will find it advantageous to preempt.

Credibility depends on precommitments by both sides to implement a (probabilistic) threat. Such precommitments, backed up by the formidable second-strike capability of the superpowers' largely invulnerable submarine-launched missiles, certainly seem to characterize the nuclear retaliatory policies of the superpowers. As we indicated earlier, however, probabilistic threats of full-fledged retaliation may be interpreted as diminished responses to a provocation, but carried out with certainty.

Such responses in repeated play of a game would, it seems, drive one up the escalation ladder.[12] Fortunately, the nuclear rung has never been reached in any superpower confrontation, which seems at least partially explained by the dynamic stability of the deterrence equilibrium: After any perturbation in a player's preemption probability, that probability tends to be restored to zero. Thus, equilibrium is maintained by the powerful force of fixed retaliation probabilities.

This self-restoring equality of the deterrence equilibrium will be reinforced by robust threats, which are always above the threshold level necessary to deter but never commit a player to certain retaliation. Because these threats are both equally damaging and equally costly whatever one side thinks the other might do, they would, we believe, enhance the stability of the deterrence equilibrium in a game of incomplete information.

Perhaps the most difficult question to answer is: What, operationally, constitutes a robust threat? We argued earlier that the present nuclear doctrines of the superpowers seem to preclude a certain response except, perhaps, to a massive nuclear attack wherein all signs are unambiguous. On the other hand, they would seem to imply probabilities above the (minimal) threshold values. But are these threats, and the actions to make them credible, as nonprovocative as practicable?

If false signals should trigger an unprovoked attack, the consequences surely would be deadly. It therefore seems better to err on the side of not being responsive enough—having "only" a probabilistic threat, which our model indicates is quite sufficient if a doomsday machine largely beyond human control undergirds it—rather than make one's retaliation too automatic or too sensitive to provocation.

Emphatically, we are not suggesting by our usage of "probabilistic threats" that the throw of dice or the spin of a roulette wheel determines, or should determine, whether an American president or a Soviet general secretary retaliates against a first strike. Rather, the uncertainties of retaliation are already inherent in the imperfect performance of any C^3I system. It can presumably be designed, nonetheless, to be more responsive or less responsive, as circumstances seem to require.

If it is hard to say exactly what constitutes a robust threat today, there is no ambiguity in our model about the undesirability of preemption. It is *never* optimal unless one can rest assured that the other side will never retaliate. Since this presumption seems hopelessly ingenuous, there seems no good reason ever to contemplate preemption, given at least threshold threats of retaliation by both sides.

Yet this is not necessarily to commend a policy of "no first use" of nuclear weapons at levels below that of superpower confrontation. In response to conventional attacks, it is conceivable that holding out the possibility of introducing nuclear weapons into a conventional conflict may help to deter an attack in the first place. But then this benefit must be weighed against the increased risk of nuclear escalation should the attack actually occur and there is no self-imposed restraint on the first use of nuclear weapons.

This and other instances of potentially appalling conflict that deterrence may prevent seem capable of game-theoretic modeling. Insofar as the Deterrence Game models non-nuclear as well as nuclear conflict, the effects of threats that underlie deterrence would seem salutary. But when threats themselves become provocative and severely undermine trust, one must ask whether their deterrent value outweighs the cost of creating an inflammatory situation.

APPENDIX

In this appendix we begin by conducting an exhaustive search for the Nash equilibria in the Deterrence Game; then we analyze their properties. The game is described is section 3.3 and depicted in figure 3.2.

$$E_R(s,q;t,p) = str_3 + (1-s)tp + s(1-t)qr_2 \qquad (1)$$
$$E_C(t,p;s,q) = stc_3 + s(1-t)q + (1-s)tpc_2 \qquad (2)$$

Our search will be broken down according to the values of s and t at the equilibrium.

Case 1: $s = t = 1$

From (1), if $t = 1$, then $\partial E_R/\partial s = r_3 - p$. Since $s = 1$ at equilibrium only if $\partial E_R/\partial s \geqq 0$, $p \leqq r_3$ is necessary. Analogous consideration of (2) shows that $q \leqq c_3$ at any equilibrium with $s = 1$. Now suppose that $t = 1$ and $p \leqq r_3$. From (1), Row's expected payoff is

$$E_R(s,q;1,p) = p + s(r_3-p)$$

so that Row can never do better than to choose $s = 1$ and $q \leq c_3$. Similarly, $t = 1$, $p \leq r_3$, is Column's best response to $s = 1$, $q \leq c_3$.

Therefore, the only equilibria consistent with case 1 are

$$s = 1, q \leq c_3; \qquad t = 1, p \leq r_3 \tag{3}$$

The family (3) is called the *deterrence equilibrium*, since every strategy combination in the family leads to the same outcome: the cooperative outcome of the underlying Chicken game, with payoffs (r_3, c_3). Properties of the deterrence equilibrium will be adduced below.

Case 2: $t = 0$

From (1), if $t = 0$, then

$$E_R(s, q; 0, p) = sqr_2$$

so that Row can maximize his expected value only by choosing $s = q = 1$. If $s = q = 1$, then (2) shows that

$$E_C(t, p; 1, 1) = 1 - t(1 - c_3)$$

so that Column's best choice is $t = 0$, and his payoff does not depend on p.

Therefore, the only equilibria consistent with case 2 are

$$s = 1, q = 1; \qquad t = 0, p \text{ arbitrary}$$

which we call the *preemption by column equilibrium*. At this equilibrium, the outcome of the Deterrence Game is always the outcome of the underlying Chicken game associated with a "win" for Column: the outcome with payoffs $(r_2, 1)$.

Case 3: $s = 0$

This case is analogous to case 2; it reduces to the *preemption by row equilibrium*:

$$s = 0, q \text{ arbitrary}; \qquad t = 1, p = 1$$

The outcome corresponds, in Chicken, to a "win" for Row and has payoffs $(1, c_2)$.

Case 4: $0 < s < 1, \qquad t = 1$

If $0<s<1$, (2) implies that $\partial E_C/\partial p = (1-s)tc_2>0$ provided $t>0$. Thus, at any equilibrium with $0<s<1$ and $t=1$, $p=1$ also since E_C is increasing in p. Now if $t=1$ and $p=1$,

$$E_R(s,q;1,1) = 1-s(1-r_3)$$

by (1), so that Row's expected payoff is maximized only when $s=0$. This contradiction shows that there are no equilibria consistent with case 4.

Case 5: $s=1$, $0<t<1$

This case contains no equilibria, by an argument analogous to that for case 4.

Case 6: $0<s<1$, $0<t<1$

Equation (1) shows that $\partial E_R/\partial q = s(1-t)r_2$, so that for an equilibrium with $0<s<1$ and $0<t<1$, $q=1$ is a necessary condition since E_R is increasing in q. Analogously, so is $p=1$. Now suppose that $0<t<1$ and $p=1$ are fixed. To maximize

$$E_R(s,q;t,1) = t + s[tr_3-t+(1-t)qr_2]$$

it is clear that Row must choose either $s=0$ or $s>0$ and $q=1$. We discard $s=0$ since it is not consistent with case 6. Now in order that some s satisfying $0<s<1$ maximize E_R, it must be that $\partial E_R/\partial s = 0$, i.e.,

$$tr_3-t+(1-t)r_2 = 0$$

This equation implies that

$$t = t^* = \frac{r_2}{1-r_3+r_2}$$

Note that $0<t^*<1$. Analogously, for fixed s and q satisfying $0<s<1$ and $q=1$, $p=1$, and t satisfying $0<t<1$ maximize E_C only if

$$s = s^* = \frac{c_2}{1-c_3+c_2}$$

where, again, $0<s^*<1$. Finally, one can verify directly that

$$s = s^*, q = 1; \qquad t = t^*, p = 1 \tag{4}$$

is an equilibrium. We refer to this equilibrium as the *naive equilibrium*.

It is easy to show that at the naive equilibrium the players' expected payoffs are

$$E_R^* = \frac{r_2}{1-r_3+r_2}, \qquad E_C^* = \frac{c_2}{1-c_3-c_2}$$

and that $r_2 < E_R^* < r_3$ and $c_2 < E_C^* < c_3$. Thus, the deterrence equilibrium (3), with payoffs (r_3, c_3), is Pareto-superior to the naive equilibrium (4).

The deterrence equilibrium possesses a dynamic stability property that, once it forms, will (in repeated play) contribute to its persistence. To see this, assume that the deterrence equilibrium (3) has become established and, further, that

$$0 < q < c_3, \qquad 0 < p < r_3 \tag{5}$$

holds. Suppose that Column is concerned that there is some chance that Row will preempt, i.e., that $s < 1$, and that Column is therefore contemplating whether he should preempt with some positive probability. In other words, Column is no longer sure that $s = 1$ and is reconsidering his choice of $t = 1$. But now differentiation of (2) yields

$$\frac{\partial E_C}{\partial t} = s(c_3 - q) + (1-s)pc_2 \tag{6}$$

so that if (5) holds, $\partial E_C / \partial t > 0$ for every value of s satisfying $0 \leq s \leq 1$. Therefore, Column is motivated to choose $t = 1$ despite his doubts about the value of s, since E_C is increasing in t. A similar calculation shows that Row is motivated to choose $s = 1$ regardless of his perception of the value of t, providing (5) holds. Thus, probabilistic threats of retaliation that are more than minimal ($q = c_3$, $p = r_3$) but less than certain ($p = 0$, $q = 0$) will tend to restore the deterrence equilibrium if it is perturbed.

Finally, we prove that the choice of $s = 1$, $q = c_3 - \varepsilon$ is the most cost-effective way for Row to trigger a trajectory from PE_C to DE. From (6) it follows immediately that if $p = p_0$, $\partial E_C / \partial t > 0$ iff one of the following two conditions is met:

(i) $q < c_3$

or

(ii) $p_0 > 0$, $q \geq c_3$, and $s < \dfrac{p_0 c_2}{p_0 c_2 + q - c_3}$

In other words, given that either (i) or (ii) holds, it would be rational for Column to choose $t = 1$ because by so doing he would maximize his expected payoff E_C.

We will now demonstrate that the temporary cost to Row of inducing Column to choose $t = 1$ is always greater than $(1-c_3)r_2$ by showing that the initial payoff to Row must be less than c_3r_2. That is, we will prove that before Column is induced to switch from $t = 0$ to $t = 1$, Row's expected payoff [see (1)],

$$E_R(s,q;0,p) = sqr_2$$

will always be less than c_3r_2.

First, because $q < c_3$ under condition (i),

$$sqr_2 \leqq qr_2 < c_3r_2$$

Under condition (ii),

$$sq < \frac{qp_0c_2}{p_0c_2 + q - c_3}$$

The right side of this inequality has a maximum, for values of q satisfying $c_3 \leqq q \leqq 1$ (as assumed), at $q = c_3$. Hence,

$$sq < \frac{p_0c_2c_3}{p_0c_2 + c_3 - c_3} = c_3$$

and necessarily $sqr_2 < c_3r_2$. Thus, Row's expected payoff will drop by more than $r_2 - c_3r_2$ when he chooses any new strategy that induces Column to switch from $t = 0$ to $t = 1$.

This demonstrates that no possible trajectory from PE_C to DE has a smaller temporary cost than that which Row induces by choosing $s = 1$, $q = c_3 - \varepsilon$. To ensure that this trajectory terminates at DE rather than continuing to PE_R, Column would have to choose a nonretaliation probability of $p < r_3$, thereby making it unprofitable for Row to preempt him in turn at (r_3, c_3).

NOTES

This chapter is drawn in part from Steven J. Brams and D. Marc Kilgour, Optimal deterrence, *Social Philosophy & Policy* 3, no. 1 (Autumn 1985): 118–135, which is reprinted in *Nuclear Rights/Nuclear Wrongs*, ed. Ellen Frankel Paul et al. (Oxford, UK: Basil Blackwell, 1986), pp. 118–135, and in *Naturalism and Rationality*, ed. Newton Garver and Peter H. Hare (Buffalo, NY: Prometheus, 1986), pp. 241–262;

and in part from Brams and Kilgour, The path to stable deterrence, in *Dynamic Models of International Conflict*, Urs Luterbacher and Michael D. Ward (eds.) (Boulder, CO: Lynne Rienner, 1987), pp. 107–122. Reprinted here with permission from Basil Blackwell and Lynne Rienner.

1 Brams (1985b, chs. 1 and 2) and Zagare (1987). Recently, games of incomplete information have been used to model different aspects of deterrence; see, for example, Güth (1985, 1986), Powell (1987), and Morrow (1987). The deterrence of conventional conflict is implicit in Mishal, Schmeidler, and Sened (1987), in which the optimal strategies of the players depend on their perceptions of the "types" of their opponent. An excellent and up-to-date overview of game-theoretic approaches to deterrence can be found in O'Neill (1988). Such approaches are usefully contrasted with psychological approaches, as discussed, for example in Jervis, Lebow, and Stein (1985). Doubtless, psychological factors affect players' perceptions of the enemy, the enemy's threats, and, more generally, how decision makers act in a crisis. In our view, however, there is no reason in principle why these factors cannot be incorporated into game-theoretic models that allow for incomplete or imperfect information, the possibility of threats, restrictions on strategy choices that the time pressure of a crisis imposes, and the like. With the inclusion of these factors, game theory is enriched; psychology benefits from being developed within a more parsimonious and rigorous framework. Thus, we believe, the approaches are complementary; it behooves psychologists and applied game theorists to consider how insights from the two fields can better be melded.

2 Brams and Hessel (1984); the original distinction between compellent and deterrent threats is due to Schelling (1966). Recently, Petersen (1986) operationalized and tested the use of compellent and deterrent threat strategies in 135 crises, finding that decision makers were generally rational in their use of these strategies.

3 The validity of the symmetry condition in the context of Soviet-American conflict is supported by the following statement of an authority on Soviet defense policy: "The answers [to the problem posed by nuclear war and nuclear weapons] the Soviet leaders have arrived at are not very different from those given by Western governments. . . . The Soviet Union has not been able to escape from the threat of nuclear annihilation. Its leaders and its people share our predicament." [Holloway, 1983.]

4 For debate on this point, see Zagare (1985; 1987, ch. 1), Brams and Hessel (1984), and Brams (1985b, ch. 1).

5 See also Shubik (1982, pp. 265–270).

6 Schelling (1960, ch. 9). The Soviets, apparently, regard the "pressure to preempt" as an American theoretical construct of no practical relevance, given the secure second-strike capabilities that both sides possess. See MccGwire (1987, p. 274).

7 We underscore "corresponding" because equilibrium strategies in the Deterrence Game are not *interchangeable*; specifically, one player's choice of his deterrence equilibrium strategy combined with the other player's choice of his preemption equilibrium strategy do not constitute an equilibrium. See Luce and Raiffa (1957, p. 160).

8 Technically, there are infinitely many such equilibria, one for each value of p_0. This nonretaliation probability of Column is irrelevant in the subsequent dynamic analysis; however, p_0 must be adjusted, if necessary, to fall below a particular threshold if the trajectory that Row triggers by his unilateral departure is to be held on course—through subsequent deterrence of Row by Column—as noted in the text below.

9 Brams (1985b, ch. 1). A more general justification of probabilistic threats, and robust threats in particular, is given in this work. These threats are contrasted with deterministic threats; the latter are shown to strain credibility—even if they could be achieved, which we argue later is in fact quite impossible. The deterrent value of nuclear threats tinged with uncertainty was recognized more than 40 years ago by Brodie (1946, p. 74): "The threat of retaliation does not have to be 100 percent certain; it is sufficient if there is a good chance of it, or if there is a belief that there is a good chance of it. The prediction is more important than the fact."

10 For details, see Bracken (1983); Blair (1985); and Ford (1985). Gauthier (1984) claims that such precommitments are not necessary to deter aggression, but threats that are not credible are empty, and empty threats invite attack. His calculus of deterrence, we believe, is sensible only when his retaliator's threats will, because of precommitments, be implemented with a sufficiently high probability.

11 Brams (1985b, pp. 45–46). Factors that affect credibility are modeled in Cioffi-Revilla (1983) and Maoz (1983). A description of factors that contribute to the unreliability of a first strike is given in Bunn and Tsipis (1983). Uncertainty, it is important to note, may strengthen rather than undermine nuclear deterrence: "The ambiguity over command of nuclear weapons may actually contribute to the credibility of the NATO deterrent, since it makes it all but impossible to predict the outcome of a crisis that involves the alerting of military forces" (Bracken, 1983, p. 172). On this point, see the exchange of letters between Brams (1984) and Zraket (1984) and also Snow (1983) and Altfeld (1985). The value of ambiguity about not only the use but even the existence of nuclear weapons is much debated in Israel; see Friedman (1986).

12 Kahn (1965) illustrates possible sequences in full and lurid detail.

4

Winding Down

4.1 INTRODUCTION

In this chapter we modify the Deescalation and Deterrence Games to allow for the possibility of winding down from an arms race or from a failure of deterrence. Although some of the analysis of this chapter is also pertinent to the building of a strategic defense system, we defer until chapter 5 a discussion of a more fundamental modification in the Deterrence Game to take account of strategic defenses that one or both superpowers might build.

The literature on war termination is sparse, as Pillar (1983) recently indicated; earlier studies by Stein (1975) and Iklé (1971) reached a similar conclusion. Perhaps one reason for the paucity of research on war termination is the difficulty of the problem. As Schelling (1966, p. 263; italics in original) ruefully observed, "Starting a major war is about the most demanding enterprise that a planner can face . . . [but] *terminating* a major war would be incomparably more challenging." The intellectual challenge of understanding why some wars persist (e.g., the devastating conflict between Iran and Iraq that began in 1980) and why others grind to a halt or simply peter out seems equally formidable.

Generally speaking, the extant literature offers empirical generalizations from a number of case studies but few analytic results. To be sure, Pillar (1983) introduced some bargaining theory into his analysis, and Wittman (1979) provided an informative analysis based on expected utility theory, formulating the problem of reaching a settlement between warring parties as a cooperative game in a brief appendix to his article. Expected utility models have been extensively developed and tested by Bueno de Mesquita (1981, 1985) to analyze the outbreak of wars,[1] which can be looked at as the other side of the coin of war termination.[2]

Possible war termination between nuclear powers has not been extensively studied. A number of analysts have recognized, however, that providing for war termination, or enhancing so-called confidence-building measures[3] before war erupts, may degrade deterrence.[4] In fact, Abt proposed a number of practical measures for dealing with the problem of diminished deterrence that a policy of war termination may create. For example, he suggested that the United States forgo nuclear attacks on Soviet cities and instead launch a "retaliatory invasion" against the Soviet Union if deterrence fails (Abt, 1985). But we know of no attempts to model, within a game-theoretic framework, possible trade-offs between deterrence and war-termination procedures that are designed to reduce the costs of war should it occur.

To some analysts, the failure of nuclear deterrence between the superpowers would almost surely be catastrophic, perhaps resulting in a "nuclear winter." Believing that there would be virtually no hope of escape once nuclear weapons were introduced, even on a limited basis, into a conflict, these analysts argue that the termination of a nuclear war is, therefore, a moot question. On the other hand, other analysts contend that the very idea that a limited nuclear exchange might escalate into a nuclear holocaust is so horrendous that both sides would be induced to terminate an incipient nuclear conflict as rapidly as possible.

In this chapter, we do not take sides on the likelihood of these events but instead address a related issue: *Is winding down a limited nuclear war consistent with deterring it in the first place?* If the answer to this question is no—one cannot maintain a credible deterrent and still hold out the prospect of cooperating if it fails—then a nuclear power faces a serious dilemma. For if it relaxes its deterrent posture by saying it could contemplate a return to the *status quo ante*, it may invite attack. But if it resolutely threatens an opponent with a terrible outcome should deterrence fail, and failure indeed occurs, than an ominous situation is likely to turn into an utter calamity.

We analyze this dilemma by changing the assumption of the Deterrence Game that, once an attack has occurred, both sides can entertain only a move to the final disastrous outcome in the Deterrence Game. Instead, we assume the possibility that both sides can revert with some exogenous probability to the initial cooperative outcome (before deterrence failed) and thereby escape the disaster and, hence, the costs of a failure in deterrence. (Later we make an analogous change in the rules of the Deescalation Game.)

Our main result in this new game, called the Winding-Down Game, is that the probability of reversion can never be 1 if both sides are, by means of their probabilistic threats of retaliation, to make it irrational for the other side to attack in the first place. In other words, rational deterrence is inconsistent with a rational promise to wind down and return to cooperation, with certainty, if deterrence fails. Intuitively, this is so because if one side makes it "too easy" for its opponent to extricate itself from a possibly cataclysmic situation, it makes the threat of retaliation and its possibly ghastly consequences no longer so fearsome. Such a degradation in fear may encourage preemption.

Fortunately, some allowance for winding down *is* consistent with rational deterrence—that is, one can build in *some* prospect of an escape from disaster—but the players must threaten more certain retaliation if an opponent preempts. However, if deterrence fails, one cannot raise the expected payoffs of the players "too much" without undermining the dominant-strategy equilibrium property of mutual cooperation (the deterrence equilibrium) in the Winding-Down Game. This is so because if the probability of an escape from the *failure* of deterrence is too high, the disastrous outcome will be made too attractive (i.e., better than being preempted), in which case the underlying game is transformed from Chicken into Prisoners' Dilemma, wherein the cooperative strategies can never be made dominant.

We next extend the analysis to arms races, modeled by the Deescalation Game based on Prisoners' Dilemma. When winding down is incorporated into this game to create the Arms Reduction Game, mutual cooperation can always be rendered stable through more probable threats of retaliation—even when the probability of winding down is 1—but this outcome (the deescalation equilibrium) does not have the dominant-strategy equilibrium property.

We conclude with a discussion of the trade-off between using threats to deter preemptive attacks and deescalate arms races, on the one hand, and being willing to heal wounds after preemption or escalation has occurred on the other. Implications of these findings, especially for encouraging mutual cooperation between the superpowers, are discussed.

4.2 THE WINDING-DOWN GAME

There are two ways that deterrence can fail in the Deterrence Game (figure 3.2): Row and Column may choose $(0,0)$ initially [with probability $(1-s)(1-t)$], or they may reach it after one player preempts and the other retaliates [with probability $(1-s)t(1-p) + s(1-t)(1-q)$]. However this disastrous outcome arises, assume that with probability w (for winding down) the players can escape it and return to (r_3, c_3), while with probability $(1-w)$ they remain stuck at $(0,0)$. Thus, their expected payoffs at the disastrous outcome become

$$w(r_3, c_3) + (1-w)(0,0) = (wr_3, wc_3)$$

Call the resulting game (with winding down) the New Deterrence Game: it duplicates the figure 3.2 game except for (wr_3, wc_3) in the lower right-hand entry.

It is convenient to normalize the payofffs of the players in the New

Deterrence Game. By subtracting wr_3 from Row's payoffs and then dividing the result by $1 - wr_3$ [this makes Row's worst payoff (lower right entry) 0, and his best payoff with no retaliation by Column (lower left entry) 1], we obtain the following formula for the new payoffs of Row in the normalized game:

$$\text{New } r_i = \frac{\text{Old } r_i - wr_3}{1 - wr_3}$$

A similar transformation can be used to normalize Column's payoffs.

Call the New Deterrence Game with normalized payoffs the Winding-Down Game, and distinguish its payoffs by primes. Thus for Row,

$$r_1' = 0; \qquad r_2' = \frac{r_2 - wr_3}{1 - wr_3}; \qquad r_3' = \frac{r_3 - wr_3}{1 - wr_3}; \qquad r_4' = 1$$

This game is shown in figure 4.1 with probabilities of nonretaliation incorporated in the payoffs, as in the Deterrence Game. Observe that when $w = 0$, $r_2' = r_2$ and $r_3' = r_3$, which would make the Winding-Down Game the same as the figure 3.2 Deterrence Game.

It should be noted that the Winding-Down Game is strategically equivalent to Deterrence Game (with r_2, r_3, c_2, and c_3 replaced by r_2', r_3', c_2', and c_3'), and that the analysis of the Deterrence Game therefore applies equally well to the Winding-Down Game. One proviso should be noted: The above transformation assumes that $w < r_2/r_3$ (in order that $r_2' > 0$), and a similar condition for Column. This analysis thus applies to small values of w. The case of large values of w will be treated in section 4.3.

The deterrence equilibrium and the preemption equilibria of the Deterrence Game have their counterparts in the Winding-Down

		Column	
		t	$1-t$
Row	s	(r_3', c_3')	(qr_2', q)
	$1-s$	(p, pc_2')	$(0,0)$

Key: (r_i', c_j') = (payoff to Row, payoff to Column)
r_3', c_3' = next best; r_2', c_2' = next worst
s, t = probabilities of nonpreemption
p, q = probabilities of nonretaliation
Normalization: $0 < r_2' < r_3' < 1$; $0 < c_2' < c_3' < 1$

Figure 4.1 Winding-Down Game.

Game, except that r_3 and c_3 become r'_3 and c'_3 in the Winding-Down Game. Thus in the Winding-Down Game, Row will be deterred from preempting Column if Column's probability of nonretaliation, p, satisfies

$$p \leq r'_3 = \frac{r_3 - r_3 w}{1 - r_3 w}$$

Note that $r'_3 < r_3$ because

$$r_3 - r'_3 = \frac{r_3 - r_3^2 w - r_3 + r_3 w}{1 - r_3 w}$$

$$= \frac{r_3 w (1 - r_3)}{1 - r_3 w} > 0$$

Therefore, because

$$p \leq r'_3 < r_3$$

Column's threat of retaliation $(1 - p)$ in the Winding-Down Game must be greater than this threat in the Deterrence Game to deter Row from preempting.

Call the threshold probability at which Row would be indifferent between preempting or not p_t:

$$p_t = r'_3 = \frac{r_3 - r_3 w}{1 - r_3 w} \tag{4.1}$$

In figure 4.2, p_t is plotted as a function of w. Its slope,

$$\frac{dp_t}{dw} = \frac{-r_3(1 - r_3)}{(1 - r_3 w)^2}$$

is negative; the function is concave in w because

$$\frac{d^2 p}{dw^2} = \frac{-2r_3^2(1 - r_3)}{(1 - r_3 w)^2} < 0$$

This means that p_t falls at an increasing rate as w increases. Thus, as the probability w of winding down rises, the nonretaliation probability p_t needed to sustain deterrence at the threshold falls even more rapidly. Hence, as w increases one must become increasingly more threatening. In other words, an increasing probability w of escaping

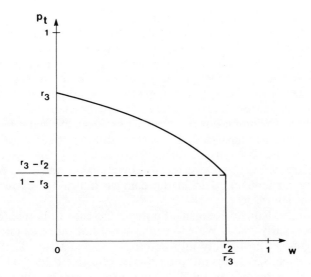

Figure 4.2 Threshold probability of nonretaliation (p_t) as a function of probability of winding down (w) in Winding-Down Game.

from disaster requires an even faster-increasing threshold retaliation probability $1 - p_t$ to ensure that the failure of deterrence will remain so costly to the preemptor that he will *not* preempt.

It is worth reiterating (see section 3.5) that we are not suggesting that players select retaliation probabilities according to some random device. Rather, the policies and operational procedures they choose, such as launch-on-warning or the use of particular safeguards against accidental nuclear war, bear on the certainty, or lack thereof, of their responses to different kinds of provocations. The uncertainties inherent in C^3I create what we called in section 3.4 a PDM (probabilistic doomsday machine) that helps to ensure the success of deterrence.

Increasing the retaliation probability to offset the winding-down probability may be accomplished by means other than creating a more hair-trigger response. Russett (1983, pp, 148–153), for example, has argued that changing the *nature* of one's retaliatory threat—specifically, through "counter-combatant targeting"—may have the effect of raising its expected damage without raising the probability of escalating to countervalue strikes, which could result in a nuclear winter.

In Presidential Directive 59 (PD-59, issued on July 25, 1980), Jimmy Carter authorized the targeting of the Soviet political leadership in a so-called decapitation strike, presumably because he and his advisors saw the Soviets as placing a higher value on this target than others previously selected. If such a change in the nature of the threat is in fact perceived by an opponent as a more devastating blow, one's *probability* of retaliation can remain unchanged and still counterbalance an increased probability of winding down—should the blow be struck—in order to maintain the same level of deterrence.

In terms of our model, such a greater threat by both players would lower the value of the original (r_1, c_1) just enough that, because the possibility of winding down mixes in (wr_3, wc_3), the expected value of disaster remains unaffected. To be sure, if the probability of the threat does not increase, its fearsomeness does, so the winding-down probability still requires enhanced retaliation—if not in its probability of being carried out, then in the damage inflicted if carried out—to preserve deterrence.

Curiously, Ronald Reagan's Strategic Defense Initiative (SDI)—or, more popularly, "Star Wars"—can also be thought of as introducing a winding-down probability into deterrence calculations. The reasoning is as follows: SDI holds out the promise of a shield that can provide a "national technical means," in the vernacular of the verification literature (see chapter 8), for winding down. That is, if SDI proves feasible, each superpower, by building its own Star Wars defense, can attenuate disaster by itself; it does not require the cooperation of the other superpower.

In game-theoretic terms, SDI can thus be viewed as a means, in our noncooperative game-theoretic framework, to wind down a preemptive strike, with the probability of winding down dependent on the magnitude of this strike and the quality of one's defenses. We will come back to this point in section 4.4.

To return to the model, at $w = 0$, when there is no possibility of escaping disaster, $p_t = r_3$, as in the Deterrence Game. As w increases, the game underlying the Winding-Down Game remains Chicken—and the results for the Deterrence Game described in section 3.3 can be applied—as long as the next-worst payoff of the Winding-Down Game, r_2', is greater than the worst payoff, $r_1' = 0$, or $w < r_2/r_3$. Analogous consideration of Column's payoffs also requires that $w < c_2/c_3$.

Assuming for simplicity that $r_2/r_3 < c_2/c_3$, p_t is defined for w between 0 and r_2/r_3, as shown in figure 4.2. Moreover, since $w < r_2/r_3$, the definition of p_t, given by (4.1), yields

$$p_t > \frac{r_3 - r_2}{1 - r_2}$$

in the Winding-Down Game.

It is important to stress that p_t is the threshold value at which Row, for w in the interval $0 \leq w < r_2/r_3$, would be indifferent between preempting and not preempting. Values of p below p_t, at which

Column threatens a greater probability of retaliation, $1 - p$, would also deter Row, but these threats become more incredible as $(1 - p) \to 1$ (or $p \to 0$)—that is, as retaliation becomes certain. As shown by Brams (1985b, pp. 13–19), there is a trade-off between the effectiveness of a threat (as measured by the damage it does to the threatened party if carried out) and its incredibility (the damage it does to the threatener); robust threats (defined in section 3.4), which fall in between the threshold and the maximal values, and whose damage is independent of the strategy choices of the players, would seem to offer an optimal trade-off (Brams, 1985b, pp. 36–43). In related models of deterrence in chapters 6 and 7, we show how threats can be tailored to the level of provocation and also used to stabilize crises.

The equilibrium in the Winding-Down Game analogous to the deterrence equilibrium in the Deterrence Game has the dominant-strategy Nash equilibrium property, provided retaliation probabilities are fixed. It would therefore seem an eminently rational choice for the players, although it requires sterner threats (i.e., threats having a greater probability of being carried out) than those in the corresponding Deterrence Game that does not permit winding down after disaster strikes. Escape is also possible in the Deescalation Game based on Prisoners' Dilemma (to be analyzed next); however, stabilizing the cooperative outcome in this game with winding down—by introducing more severe threats—cannot make it a dominant-strategy equilibrium.

4.3 THE ARMS REDUCTION GAME

If r_2 is interchanged with r_1, and c_2 with c_1, in the Deterrence Game, one gets the Deescalation Game analyzed in chapter 2. The question we address in this section is whether, if winding down from (r_2,c_2) to (r_3,c_3) may occur with some probability w in this game (figure 2.2), the stability of the deescalation equilibrium can still be preserved. Put differently: Is allowing for the possibility of escaping escalation, once there, consistent with preventing it in the first place?

We proceed in the same manner as in section 4.2, first showing how the Deescalation Game can be transformed into a new game, called the Arms Reduction Game, when w is introduced as a parameter in the Deescalation Game. The reason we use the terminology of "arms reduction" is to suggest that arms races may be ameliorated by introducing a probability of winding down into our Deescalation Game model. The crucial question this new element poses is: Is the stability of (r_3,c_3) jeopardized in this game?

If, once the players in the Deescalation Game reach (r_2,c_2), they can move to (r_3,c_3) with probability w but remain stuck at (r_2,c_2) with probability $1 - w$, their expected payoffs at the noncooperative position become

$$w(r_3,c_3) + (1-w)\,(r_2,c_2) = (wr_3 + [1-w]r_2,\; wc_3 + [1-w]c_2)$$

in the new game to which winding down gives rise. Because the payoffs in this game continue to range between $r_4 = 1$ and $r_1 = 0$, no normalization is necessary: the payoffs (r_i',c_i') in the Arms Reduction Game duplicate those in the Deescalation Game, except for those in the lower right-hand entry,

$$(r_2',c_2') = (wr_3 + [1-w]r_2,\; wc_3 + [1-w]c_2)$$

Note that because the payoffs on the right side of this equation are simply a convex combination of the intermediate payoffs in the Deescalation Game, they remain generally the next-worst payoffs. More specifically, for Row, $r_2 \le r_2' < r_3 = r_3'$ except when $w = 1$; then (r_2',c_2') is the same as the deescalatory $(r_3',\, c_3') = (r_3,c_3)$ outcome for both players.

Because the Arms Reduction Game is strategically equivalent to the Deescalation Game, the same reasoning applies. From the analysis in chapter 2, it follows in the Arms Reduction Game that Row will be deterred from escalation if Column's probability of nonretaliation, p, satisfies

$$p \le \frac{r_3' - r_2'}{1 - r_2'} = \frac{r_3 - [wr_3 + (1-w)r_2]}{1 - [wr_3 + (1-w)r_2]} = \frac{(1-w)\,(r_3 - r_2)}{1 - r_2 - w(r_3 - r_2)}$$

[The right side of this inequality is a well-defined probability since

$$1 - r_2 - w(r_3 - r_2) \ge 1 - r_2 - (r_3 - r_2) = 1 - r_3 > 0$$

and since

$$(1-w)\,(r_3 - r_2) < 1 - r_2 - w(r_3 - r_2)$$

because $(r_3 - r_2) < (1 - r_2)$.] As in the Winding-Down Game, let the threshold probability in the Arms Reduction Game at which Row would be indifferent between escalating or not escalating be

$$p_t = \frac{(1-w)\,(r_3 - r_2)}{1 - r_2 - w(r_3 - r_2)}$$

In figure 4.3, p_t is plotted as a function of w. Its slope,

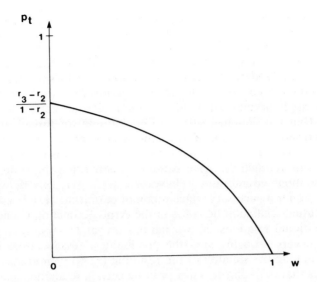

Figure 4.3 Threshold probability of nonretaliation (p_t) as a function of probability of winding down (w) in Arms Reduction Game.

$$\frac{dp_t}{dw} = \frac{-(1 - r_3)\,(r_3 - r_2)}{[1 - r_2 - w\,(r_3 - r_2)]^2}$$

is negative, and its second derivative,

$$\frac{d^2 p_t}{dw^2} = \frac{-2(1 - r_3)\,(r_3 - r_2)^2}{[1 - r_2 - w\,(r_3 - r_2)]^3}$$

is negative also.

Thus, p_t is concave in w, as in the Winding-Down game. This means that, in the Arms Reduction Game, an increasing probability w of escaping from the mutual-escalation outcome requires a faster-increasing threshold retaliation probability $1 - p_t$ to ensure that escalation from (r_3, c_3) will not be profitable.

At $w = 0$, when there is no possibility of winding down from mutual escalation,

$$p_t = \frac{r_3 - r_2}{1 - r_3}$$

as in the Deescalation Game. Unlike in the Winding-Down Game, p_t is defined for all w between 0 and 1; the Prisoners' Dilemma structure underlying the Arms Reduction Game does not depend on w.

Put another way, if $w > r_2/r_3$ and $w > c_2/c_3$ in the Winding-Down Game, the underlying game is no longer Chicken but Prisoners' Dilemma.[5] Then the results of this section for the Arms Reduction Game apply: Even when winding down is a virtual certainty, (r_3,c_3) is a Nash (but not dominant-strategy) equilibrium for threats at or above the threshold retaliation probabilities in this game.

This consequence says that escape from (r_2,c_2) is always possible in the Arms Reduction Game without undercutting the stability of the deescalation equilibrium. However, as $w \to 1$, $p_t < 0$, so certain retaliation is a necessary concomitant of permitting certain escape from the mutual-escalation outcome in the Arms Reduction Game. Thus in international relations, if two nations are at the (r_2,c_2) trap—as the superpowers probably are—the possibility of escape from it without undermining the stability of the resulting (r_3,c_3) compromise outcome implies that both players must promise certain retaliation should either subsequently depart from (r_3,c_3), and both must have the capability assuredly to detect such departures.

The escape route in the Arms Reduction Game based on Prisoners' Dilemma is therefore wider than in the Winding-Down Game based on Chicken: One can always provide for it in the former game, but not the latter, while still maintaining the stability of (r_3,c_3). Stability, though, must be bought at the price of more certain retaliation by a player whose opponent deviates from the mutually cooperative outcome. Furthermore, the probability of retaliation at the threshold value must increase faster and faster in either game as the probability of winding down increases.

4.4 CONCLUSIONS

In chapter 3 we showed that though both deterrence and preemption lead to equilibrium outcomes in the Deterrence Game, the deterrence equilibrium is, in one sense, more stable than the preemption equilibria. The reason is that, if retaliation probabilities are fixed, there is a rational path that can be taken to the deterrence equilibrium, but not from it to the preemption equilibria; however, initiating movement along this path is not without risk. This finding echoes what we found in the Deescalation Game of chapter 2: There is a rational path from the escalation equilibrium to the deescalation equilibrium, but not vice versa; but even more felicitously than in the Deterrence Game, it is costless to initiate, in part because the latter equilibrium is Pareto-superior to the former.

In this chapter we have amended the rules of the Deterrence and Deescalation Games to allow for an exogenous probability of escape if deterrence fails or if there is mutual escalation in an arms race. The principal question we sought to answer is: What effects do these escape mechanisms have on the stability of the deterrence and deescalation equilibria in each game?

Our main finding is that, in principle, winding down in either game is possible. Yet in order not to destabilize the cooperative outcomes in the Deterrence and Deescalation Games, the players must upgrade the level of their retaliatory threats against departures by their opponents from these outcomes.

In the Winding-Down Game, the probability of winding down can be increased only up to a certain point and still be offset by increasing threats that ensure the strong (dominant-strategy) equilibrium status of no preemption. Beyond this point, however, the underlying game, formerly Chicken, becomes Prisoners' Dilemma; the cooperative equilibrium of the resulting Arms Reduction Game can be sustained even as the probability of winding down from mutual escalation increases to 1, though with a somewhat weakened form of stability (not dominant-strategy).

In the Arms Reduction Game, then, the deescalation equilibrium is not a dominant-strategy Nash equilibrium, as is the deterrence equilibrium in the Winding-Down Game, when probabilities are fixed. In both games, the winding-down probability forces the retaliation probability up in order to sustain the cooperative equilibrium, and at an increasing rate. Thus, as escape becomes more and more probable, the minimum probabilistic threat of retaliation must increase still faster to deter preemption or prevent escalation.

We suggested that instead of increasing the probability of the threatened punishment's being carried out, the punishment itself might be made more painful to compensate for the decline in its deterrent value when winding down is possible. Altering the nature of the threat may be preferable to increasing its probability of execution, because the former course may lower the probability of a crisis erupting into a full-blown disaster because retaliation, if harsher, is less certain. Of course, when it occurs, its more severe consequences may make it potentially more escalatory, so the probability of a cataclysm may in the end not be reduced.

Star Wars, if it ever proves technically feasible, could provide an escape route that does not require an opponent's cooperation. But if deterrent threats cannot be abandoned entirely—as is very likely since a "leakproof" system is almost certainly an impossibility—it will require an escalation in threats, just as more dreaded threats are needed to preserve deterrence when cooperative means of winding down are introduced into the Deterrence and Deescalation Games.

Worse, as we shall show in chapter 5 in another modification of the Deterrence Game, Star Wars may have preemption-inducing qualities, especially if its development by the superpowers is unbalanced and one side gets a substantial lead over the other.

Basically, it seems, there is no "free lunch" if one wants to provide both for deterrence and a possible escape from its failure. Nalebuff has reached a similar conclusion, using a rather different model of brinkmanship: Deterrence depends on conflicts' having the potential to escalate out of control.[6]

We regard our findings with mixed feelings. At the same time that one can provide for the possibility of winding down, one must counterbalance such cooperative efforts with greater threats to heighten the grim consequences if deterrence fails or escalation occurs. The "hot line" and other communication links between the superpowers, as well as pledges to inform the other side of future military exercises and not encrypt telemetry from missile tests, may be viewed as small steps intended to defuse potential crises and decrease the probability of nuclear war (discussed in chapter 9). These measures, we think, are commendable and do not substantially undermine nuclear deterrence between the superpowers.

Attempts to reduce significantly stockpiles of strategic weapons seem to have been less successful, perhaps in part because a continuing arms race is not so terrifying as a possible failure of nuclear deterrence. But our theoretical analysis indicates that it is indeed possible to effect arms reductions and, at the same time, stabilize the cooperative outcome in the Arms Reduction Game; practically, reprisals for cheating on arms control agreements would have to be surer to discourage violations once reductions had commenced. Insofar as both sides recognize the need for a strict stand against cheating, and conclude that verification techniques are adequate for supporting such a policy (discussed in chapter 8), deescalation of the arms race would appear to be both possible and desirable. An important lesson of our analysis, however, is that deescalation cannot be safely pursued without strengthening threats of reprisal if cheating is detected.

NOTES

This chapter is drawn from Brams and Kilgour, Winding down if preemption or escalation occurs: a game-theoretic analysis, *Journal of Conflict Resolution* 31, no. 4 (December 1987). Reprinted by permission of Sage Publications, Inc.

1 Extensions of these models are developed in Maoz (1985).
2 Wittman (1979) makes this argument.

3 See Bert and Rotfeld (1986) and Borawski (1986).

4 Vick and Thompson (1985); Wieseltier (1985, esp. pp. 844–54); and Abt (1985, pp. 33–35).

5 The threshold might be exceeded for one player and not the other; for example, it might occur that $r_2/r_3 < w < c_2/c_3$. In this case, which we do not analyze here, the underlying game is one that has been dubbed "called bluff" in Snyder and Diesing (1977, pp. 107ff). This is game no. 39 in the classification of 2×2 ordinal games given in Rapoport, Guyer, and Gordon (1976).

6 Nalebuff (1986). The control of nuclear weapons and relevant historical background are discussed in Ball (1981). Lebow (1987) discusses three "sequences to war"—preemption, loss of control, and miscalculation—and proposes a number of measures to bolster crisis stability. And on a more specific point, George (1986, pp. 212–13, fn 20) discusses the idea of a joint US–USSR "crisis control center"; see also Ury (1985). There would seem, according to our analysis, to be a fundamental inconsistency between enhancing crisis management and enhancing nuclear deterrence; the specific trade-offs between these two approaches to avoiding nuclear war deserve to be carefully modeled. An up-to-date and comprehensive analysis and assessment of different aspects of nuclear management is given in Carter, Steinbruner, and Zraket (1987).

5

Star Wars

5.1 INTRODUCTION

The furor caused by Ronald Reagan's Strategic Defense Initiative (SDI)—or, more popularly, "Star Wars"—has not abated since he announced it in his speech of March 23, 1983.[1] We suggest there are two main reasons for the continuing controversy. First, the initiative took the United States and the world, including the defense community, by surprise; even today, some years later, questions about SDI's feasibility and effectiveness sharply divide proponents and opponents.

Second, an extraordinary amount of money over a short period of time has been allocated to research for and testing of various aspects of SDI; this has created a growing pro-SDI constituency in addition to instigating a strong anti-SDI reaction. Moreover, given the possibility of much more money to come for further research and development and eventual deployment of the system, the forces on both sides are likely to expand as competition for SDI funds intensifies.

Our concern in this chapter is not with this domestic political competition but rather with assessing SDI's strategic effects on nuclear deterrence. Our approach is unorthodox and will perhaps seem inordinately abstract for such a concrete issue over which so much debate has raged. We ignore all questions connected with the technical feasibility of SDI, which have dominated debate about Star Wars, and even questions of whether a viable defense against nuclear weapons is possible, much less cost-effective. Neither do we try to analyze SDI's international political effects, including its impact on Soviet decision making, the behavior of allies, or the prospects of negotiating arms control agreements. Rather, our focus is on modeling the stability of nuclear deterrence as Star Wars becomes a more and more effective defensive system.

To be sure, the "stability question" has been discussed in the literature, but only recently has there been formal modeling of this question to try to ascertain the precise trade-offs between deterrence and defense.[2] Ostensibly, as each side develops measures that could blunt a *first* strike against it and enhance its ability to retaliate, a potential opponent's uncertainties about the putative advantages of attacking first would seem to increase, thereby strengthening deterrence. On the other hand, because this same defensive system can be used to ward off or attenuate a *second* strike, it could undermine the threat of a retaliatory counterattack meant to deter the first strike. Thus, the effects of Star Wars may pull in different directions, helping and hurting deterrence at the same time. The net impact of these contradictory forces is what we try to sort out in our analysis.

We begin by modifying the rules of the Deterrence Game once again, this time by allowing not for the possibility of winding down directly but rather for the possibility of varying levels of defense against first and second strikes in a Star Wars Game. We then examine what kinds of Nash equilibria can exist in this game.

Next we posit three different scenarios for the Star Wars Game and show graphically the regions, often overlapping, in which various equilibria exist. In the first scenario, for example, we assume that each side's first-strike and second-strike defenses are the same and show that in addition to the deterrence and unilateral preemption equilibria of the Deterrence Game, new and disturbing equilibria crop up in the Star Wars Game—namely, equilibria of mutual preemption—especially as Star Wars defenses become better and better.

The general picture that emerges from the different scenarios—putting aside the question of the cost of building a Star Wars system—is not encouraging. Basically, at low levels of defense, Star War leaves mutual deterrence intact even when only one side has a Star Wars system, chiefly because deterrence via a retaliatory second strike is still possible. At higher levels, an incentive for mutual preemption develops, especially when one side is substantially ahead of the other and is therefore not deterred to the extent it would be if it were facing an equal foe; this may in turn induce the weaker side also to preempt.

A symmetry in Star Wars capabilities sometimes offers help, staving off preemption and prolonging deterrence as defenses improve. Yet at low levels of defense unilateral preemption is an equilibrium, and—more dangerous—at high levels, mutual preemption is an equilibrium and may be the only one. Fortunately, the stability of mutual preemption in the Star Wars Game is counterbalanced by the fact that deterrence equilibria often coexist with the mutual-preemption equilibria when the Star Wars capabilities of the two sides are relatively high and more or less equal, and, in most plausible situations, the deterrence equilibria are better for both players.

In a crisis, however, the notion of preemption may come to the fore and, especially if it is suspected that one's opponent may preempt, one does better by striking first (or simultaneously) than by absorbing a first strike and then retaliating with diminished capabilities. This incentive toward mutual preemption never exists in the Deterrence Game, which suggests that Star Wars, paradoxically, may be compatible with deterrence (especially at low levels of defense) yet undermine crisis stability (by making mutual preemption an equilibrium at higher levels).

After exploring these contradictory effects of Star Wars, we investigate the tricky time path that it would appear the superpowers must adhere to if they are to avoid a possible cataclysm, particularly in the transition period from principally deterrent to principally defensive postures. We conclude with recommendations on controls to strengthen crisis stability, the Achilles heel of Star Wars, if its development by the superpowers is uncoordinated and one side should get a defensive edge on the other.

5.2 THE BASIC STAR WARS GAME

Recall that in the Deterrence Game (section 3.3) Row could choose any nonpreemption probability s and nonretaliation probability q, and Column could choose nonpreemption and nonretaliation probabilities of t and p, respectively. In the Star Wars Game we restrict these choices by assuming that the players' defensive capabilities put upper bounds on the amounts of preemption $(1-s, 1-t)$ and retaliation $(1-q, 1-p)$ that their opponents can choose or, equivalently, lower bounds on s, t, q, and p.

To be more precise, we assume that the partial shield of Column's Star Wars defense implies that Row's *first* strike must satisfy $v_C \leq s \leq 1$, where $0 \leq v_C < 1$; similarly, Row's defense will force Column's first-strike choice to satisfy $v_R \leq t \leq 1$, where $0 \leq v_R < 1$. In each case, the parameters v_C and v_R constrain the range of the players' choices, giving minimum levels of cooperativeness (or maximum levels of noncooperativeness) of a first strike. Put another way, because v_C and v_R measure Column and Row's defensive capabilities against first strikes by their opponents, $1 - v_C$ and $1 - v_R$ measure their opponents' maximal attack capabilities and may be thought of as ceilings on the magnitude of the players' first strikes: Row and Column's choices of $1-s$ and $1-t$ may not exceed these ceilings $(1-s \leq 1-v_C$ and $1-t \leq 1-v_R)$. (These upper bounds are equivalent to the lower bounds of v_C for s and v_R for t given earlier.)

We use additional parameters w_C and w_R to denote the restrictions on the players' *second* strikes imposed by Column's and Row's defense capabilities. Specifically, these restrictions are

$$w_R \leq p \leq 1 \qquad \text{and} \qquad w_C \leq q \leq 1$$

where $0 \leq w_R \leq 1$ and $0 \leq w_C \leq 1$. For example, $1 - w_R$ is a ceiling on the magnitude of a second strike by Column in response to a first strike by Row.

It is reasonable to assume that the parameters w_R and w_C are functions of v_R and v_C (and, possibly, of other factors as well). Generally we would expect that w_R is an increasing function of v_R (because Row's first-strike defense would help against second strikes as well) and a decreasing function of v_C (because the better Column's defense against Row's first strike, the stronger Column's second strike can be). In fact, to illustrate specific results of our basic model in section 5.4, we postulate in one scenario

$$w_R = v_R(1 - v_C); \qquad w_C = v_C(1 - v_R)$$

Before investigating the implications of this and other functional relationships, it is useful to clarify how the Star Wars constraints, v_C, v_R, w_C, and w_R, affect the payoffs of the players and ultimately their rational strategy choices in the Star Wars Game. To begin with, assume there is no first strike by Column (i.e., $t = 1$). If Row also chooses no first strike (i.e., $s = 1$), then the payoffs to the players are (r_3, c_3). On the other hand, if Row's actions are considered to be a first strike (with probability $1 - s > 0$), and there is no Star Wars defense by Column, the payoffs to the players are $(1, c_2)$, at least temporarily (since Column might retaliate).

Now assume that Column has some defensive capability, so the maximal first strike Row can inflict on him is $s = v_C$, somewhere between the extremes of $s = 0$ and $s = 1$. This means that the payoff Row derives from a maximal first strike is equivalent to a mixture of no first strike (r_3) and a maximal first strike against no defense (1) in the proportions of v_C to $1 - v_C$. Similarly, any less-than-maximal first strike $(s \neq 0)$, mixing $s r_3$ and $1 - s$, will be attenuated in the same proportions.

Thus, the greater v_C, the more Row is forced to partake of r_3 rather than his maximal payoff of 1 after he preempts; in effect, this forces Row to be more cooperative. The reason, of course, is that Column's Star Wars defense prevents Row from moving beyond v_C toward $s = 0$. By contrast, when Column has no defense, $v_C = 0$, allowing for $s = 0$ and payoffs of $(1, c_2)$; when Column's defense is total ("leakproof") $v_C = 1$, ensuring that $s = 1$ so that Row's choice of a first strike is payoff-equivalent to the choice of no first strike, and the payoffs are (r_3, c_3). It is worth noting that the $v_C : (1 - v_C)$ payoff ratio we assume for Row when he strikes first applies to the mixture between c_3 and c_2 for Column.

Now consider possible second strikes. Assume that Column's actions are believed to constitute a first strike against Row (with probability

$1 - t > 0$). If Row does not retaliate, the players' payoffs are $(r_2, 1)$, whereas maximal retaliation by Row could give payoffs of $(0, 0)$ if Column had no defense.

If Column has some defensive capability, however, the maximal retaliation, w_C, that Row can inflict on him falls between the extremes of $q = 0$ and $q = 1$. Analogous to our assumptions about first strikes, we interpret Row's maximal second strike against Column's defense as payoff-equivalent to a mixture in the proportion $w_C : (1 - w_C)$ of no retaliation (payoff of r_2 to Row) and maximal retaliation against no defense (payoff of 0).

As with v_C in the case of first strikes, $w_C = 0$ means that Row can retaliate to any extent he chooses, and $w_C = 1$ means that, effectively, Row has no retaliation option. As before, we assume that the payoff mixture for Column is in the same ratio as for Row.

These assumptions for the players fix the payoffs they derive in the Deterrence Game when Star Wars defenses are erected. Expected payoffs in the resulting Star Wars Game are the same as those given by (3.1) and (3.2) in the Deterrence Game, except that there are now the previously given constraints on the values that the players' strategy choices—s and q for Row, t and p for Column—can assume in the Star Wars Game.

5.3 NASH EQUILIBRIA IN THE STAR WARS GAME

The Star Wars Game would be trivial if first-strike defenses were perfect $(v_R = v_C = 1)$, for then it would model two invulnerable superpowers in a defense-dominated world. On the other hand, when defenses are nonexistent $(v_R = w_R = v_C = w_C = 0)$, the Star Wars Game is identical to the Deterrence Game, wherein, as discussed earlier, two vulnerable sides can coexist in a deterrence equilibrium. Because we are interested in the problems of stability during the transition from deterrence to defense, our main concern is with the existence and properties of the deterrence equilibrium as defenses improve (v_R and v_C go from 0 to 1). We are also interested in other Nash equilibria, which might attract one or both players if there were a perturbation (random shock) to the system during the transition from a game of deterrence to a game of defense.

The Nash equilibria of the Star Wars Game are derived in the appendix. Some of these equilibria closely resemble those in the Deterrence Game, so we use the same terminology as in the latter game:

I *Deterrence equilibrium (DE)*:

$$s = 1, \quad w_C \leq q \leq c_3; \qquad t = 1, \quad w_R \leq p \leq r_3$$

As in the Deterrence Game, the players never preempt but threaten to retaliate if preempted with probabilities sufficient to deter their opponents but within their retaliation thresholds. What is crucial in the Star Wars Game is that, to make a first strike unprofitable, the players must have enough capability to retaliate effectively, despite the second-strike defenses. This requirement arises because Row must choose a retaliation probability $1 - q$ less than the ceiling of $1 - w_C$ imposed by Column's second-strike Star Wars defense (i.e., $1 - q \leq 1 - w_C$), and Column must choose an analogous probability similarly constrained by Row's second-strike Star Wars defense. As before, the payoffs to the players at DE are (r_3, c_3).

II *Preemption equilibria* (PE):

(1) PE_C: $s = 1, q = 1$; $t = v_R, w_R \leq p \leq \dfrac{v_R r_3 + (1 - v_R) r_2}{v_R}$

(2) PE_R: $t = 1, p = 1$; $s = v_C, w_C \leq q \leq \dfrac{v_C c_3 + (1 - v_C) c_2}{v_C}$

As in the Deterrence Game, the preemption equilibrium for Column (PE_C) occurs when there is maximal preemption by Column ($t = v_R$) and no preemption ($s = 1$) or retaliation ($q = 1$) by Row. Unlike the Deterrence Game, however, the p in the Star Wars Game that supports PE_C is not arbitrary but instead is constrained by the bounds given above. These say that Column's threat of retaliation, $1 - p$, if attacked (actually, he never is attacked at PE_C) will be (i) not more than that allowed by Row's second-strike Star Wars defense (which may be less than, equal to, or greater than v_R, Row's first-strike Star Wars defense), and (ii) not less than that allowed by the upper bound given above, which is a function of r_2, r_3, and v_R.[3] At PE_C, the payoffs to the players are

$$(v_R r_3 + [1 - v_R] r_2, \ 1 - v_R [1 - c_3])$$

At PE_R the payoffs are

$$(v_C c_3 + [1 - v_C] c_2, \ 1 - v_C [1 - r_3])$$

To interpret the PEs, note that the upper bound on p for PE_C is always greater than r_3, which means that Column's threat of retaliation does not have to be as great as that required to sustain DE, except when $v_R = 1$. At this extreme, Row has a perfect first-strike Star Wars defense, so preemption by Column is harmless; in fact, it is as if Column never preempted. Hence, Column's threat of retaliation must be the same as for DE to prevent Row from doing better by preempting than by capitulating. On the other hand, if Row has a less-than-perfect defense $(v_R < 1)$, the bound on Column's threat of retaliation can be relaxed from $p \le r_3$ to $p \le [(1 - v_R)/v_R]r_2$ and still ensure PE_C.

Thus, the weaker Row's first-strike Star Wars defenses are, the the less threatening Column must be to deter Row from striking simultaneously. As Row's first-strike defenses improve, Column must be more threatening to make his unilateral preemption a Nash equilibrium—that is, to ensure that preemption stays one-sided and that Column (the preemptor) is not himself preempted.

Such threat escalation may not always be possible. For example, there is an inconsistency between assuming that Row has a perfect defense against first and second strikes and assuming that Column is able to choose a p according to the conditions for PE_C. Clearly, if $v_R = w_R = 1$, the condition $1 \le p \le r_3$ for PE_C cannot be satisfied, meaning that nothing Column can do will sustain PE_C. An impregnable Row can do better by striking back unless, of course, Column is also impregnable, in which case threats of retaliation are not needed to sustain (r_3, c_3).

In fact, the no-preemption outcome can be stabilized by retaliatory threats of both players (DE), by both players' perfect first-strike Star Wars defenses, or by one player's perfect first-strike Star Wars defense coupled with a retaliatory threat by the other player that deters the first from attacking. In the last case, observe that when either $v_R = 1$ or $v_C = 1$, the above expressions for the payoffs to the players given for the preemption equilibria reduce to (r_3, c_3).

The next kind of equilibrium is not duplicated in the Deterrence Game:

III *Mutual preemption equilibrium (MPE)*:

$$s = v_C, \ q = 1; \qquad t = v_R, \ p = 1$$

Each player preempts to the maximum level permitted by the other's Star Wars defense, and neither retaliates. This equilibrium requires that $v_C \ge c^*$ and $v_R \ge r^*$, where

$$c^* = \frac{c_2}{1 - c_3 + c_2}; \qquad r^* = \frac{r_2}{1 - r_3 + r_2}$$

The payoffs to the players at this equilibrium are

$$(v_R v_C r_3 + v_R[1 - v_C] + [1 - v_R]v_C r_2, v_R v_C c_3 + [1 - v_R]v_C + v_R[1 - v_C]c_2)$$

To interpret MPE, recall that for PE_C, the reversal of the above inequality for MPE, $v_R \leq r^*$, is exactly the condition that makes the upper bound on p innocuous: Column can preempt with impunity, not even needing to threaten retaliation if attacked himself. The MPE condition, by contrast, turns this situation around: Both players' first-strike Star Wars defenses are now so strong as to blunt preemption significantly; in this case, neither player can sustain a preemption equilibrium without also threatening retaliation.[4]

Given the MPE condition, each player has a first-strike defense at or above the thresholds of c^* and r^*. With these stronger defenses, first strikes are less damaging to the parties attacked than threatening, and possibly carrying out, second strikes at a level sufficient to deter preemption by an opponent and ensure (r_3, c_3).

Each player may therefore prefer MPE to DE if the two equilibria coexist, provided his own defenses are sufficiently strong. In the appendix we indicate how, given coexistence, these preferences may create a zone near $(v_R, v_C) = (1,1)$ where MPE is Pareto-superior to DE, and always create a larger zone around $(v_R, v_C) = (0,0)$ where DE is Pareto-superior to MPE. Note that MPE and DE coincide at $v_R = v_C = 1$ in the sense that the players always receive (r_3, c_3). In other words, with a perfect first-strike Star Wars defense, nothing changes after mutual preemption (except each player's stock of arms, which is not reflected in our model), so the payoffs remain (r_3, c_3).[5]

There are two additional kinds of Nash equilibria, derived in the appendix, that we do not discuss in detail here because they are always Pareto-inferior to other Nash equilibria in the Star Wars Game. The first kind, line equilibria (LE), are similar in nature to the preemption equilibria. For Column, LE_C requires that Row's first-strike Star Wars defense parameter v_R equal exactly r^* (defined earlier), and that Column preempt maximally by choosing $t = v_R = r^*$; Row may choose any level of preemption s that satisfies $c^* \leq s$ and $v_C \leq s \leq 1$. Neither player ever retaliates (i.e., $p = q = 1$).

The main difference between MPE and LE_C is that Row, instead of preempting fully, chooses a level of preemption less than the maximum permitted by Column's first-strike defense. Although Row's level of preemption does not affect his own payoff, Column would obviously prefer less preemption (a higher s). When $s = 1$, LE_C becomes PE_C at $v_R = r^*$; this PE_C outcome gives Column a higher payoff than any LE_C, without reducing Row's payoff, so it is Pareto-superior to any LE_C outcome (with $s < 1$).

The other kind of Pareto-inferior Nash equilibrium in the Star Wars Game is analogous to the naive equilibrium (NE) in the Deterrence Game, in which both players partially preempt and neither retaliates. In the Star Wars Game, this equilibrium is characterized by $s = c^*$, $q = 1$; $t = r^*$, $p = 1$.

As in the Deterrence Game, NE in the Star Wars Game is a mixed-strategy equilibrium for both players that renders them indifferent to each other's strategy choices. It is always Pareto-inferior to DE and hence unlikely to be chosen because not only can DE, if it exists, make both players better off but also it is a dominant-strategy Nash equilibrium when a sufficient probability of retaliation is guaranteed. Moreover, it completely overlaps NE in the two most realistic scenarios to be discussed in section 5.4

The different kinds of equilibria we have described in this section may arise in all Star Wars Game, provided that the players can make strategic choices within the constraints imposed by the Star Wars defenses given for each equilibrium. We will not consider line or naive equilibria further because they are always Pareto-inferior to the other equilibria (DE, PE, or MPE) and would therefore be eschewed by rational players.[6]

5.4 PARETO-SUPERIOR EQUILIBRIA IN DIFFERENT SCENARIOS

To render more concrete the abstract definitions of the Pareto-superior equilibria, we will investigate the consequences of several different assumptions about possible relationships between the first-strike and second-strike levels of Star Wars defense: Our objective is to render more specific the discussion in section 5.3, showing interrelations among DE, PE, and MPE under different hypotheses about the nature of a strategic defense system. Altogether, we posit three scenarios, beginning with the simple case in which

Scenario 1. *First-strike and second-strike defenses are exactly the same:*

$$w_C = v_C; \quad w_R = v_R$$

In this scenario, a player's Star Wars defense is independent of whether he (say, Column) is struck first and can limit Row's first strike to $s \geq v_C$, or Column strikes first himself and then must defend against Row's retaliation, which he can limit to $q \geq w_C = v_C$. This equality assumption is sensible if it is correct to think of Star Wars as simply putting a cap on the number of missiles that can penetrate each player's defenses in either a first or second strike. Although a player's first strike before he has incurred any losses may be more devastating than his second strike after he has, the cap in either case is the same in this scenario. (In the next scenario, each player's second-strike defense is assumed to be at least as good, and generally better, than his first-strike defense.)

Figure 5.1 depicts the regions in which DE, MPE, and PE_C and PE_R exist in terms of v_C and v_R. The top graph shows overlapping regions

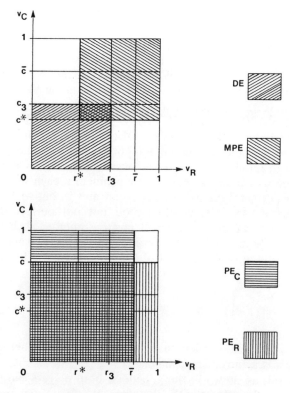

Key: v_C = level of Column's first-strike defense
v_R = level of Row's first-strike defense
DE   = deterrence equilibrium
MPE = mutual preemption equilibrium
PE_C = preemption equilibrium for Column
PE_R = preemption equilibrium for Row
Note: DE dominates MPE in their region of overlap.

Figure 5.1   Regions of existence of Nash equilibria in scenario 1:
$w_R = v_R, w_C = v_C.$

containing the former two equilibria, and the bottom graph overlapping regions containing the latter two equilibria, in the $v_R v_C$ plane. (With color overlaps, all these could be shown in one graph; in the present case, reference points are given on the two axes to indicate where the equilibria do and do not coexist.)

In the top graph, DE exists for lower levels of Star Wars defenses (up to $v_C = c_3$ and $v_R = r_3$) and MPE for higher levels (at or above $v_C = c^*$ and $v_R = r^*$). In the DE–MPE overlap region between c^* and c_3 on the vertical axis and r^* and r_3 on the horizontal axis, the DE payoff to each player is greater than the MPE payoff. This is because, as indicated in the appendix, Column prefers DE to MPE whenever they coexist and $v_C \le c_3$, and similarly for Row. Hence, DE dominates MPE whenever they overlap in this scenario.

So far, so good if the Star Wars defenses of the two players are not above the threshold values of c_3 and r_3: Excluding the PEs for now, either DE is the unique (Pareto-superior) Nash equilibrium, or it dominates the (Pareto-inferior) MPE equilibrium. However, above these thresholds MPE excludes DE, because each player's best response to the other player's preemption—if it is not accompanied by the threat of retaliation, which we consider later—is to attack. Mutual preemption would appear not to be a problem if each side's Star Wars defense is perfect, because nothing penetrates it, but what if it is not?

To illustrate the effects of a good but not perfect defense, let $v_C = v_R = 1 - \varepsilon$, where ε is a small positive number such that $v_C > c_3$ and $v_R > r_3$. Since their defenses are equal, the players are on the diagonal in the upper right part of the MPE region. After rearranging terms, Column's MPE payoff can be shown to be

$$E_C = c_3 + \varepsilon[(1 - \varepsilon) - c_3(2 - \varepsilon) + c_2(1 - \varepsilon)]$$

which may be greater or less than c_3, according to whether the quantity in brackets is positive or negative.

In general, as shown in the appendix, both players can improve on their (r_3, c_3) payoffs at DE if their Star Wars defenses are (1) sufficiently good to put them in the MPE region but (2) not perfect. This phenomenon also depends on the values of the players' utilities, as specified in the appendix and illustrated above. Nevertheless, as their defenses approach perfection, the players' payoffs invariably approach (r_3, c_3), which can be seen in the previous example by letting $\varepsilon \to 0$ in the above expression for E_C. Then $v_C = v_R \to 1$, because $v_R = v_C = 1 - \varepsilon$, and $E_C \to c_3$. Analogously, $E_R \to r_3$, so the players can actually do worse in the MPE region when they can stop *all*, rather than most, incoming missiles launched in a first (or second) strike.

It may seem astounding that mutual preemption is not only a Nash equilibrium but can also lead to higher payoffs than DE when Star Wars defenses are good but not perfect, or higher payoffs than MPE when the Star Wars defenses of both players are perfect. In effect, with good defenses the players can "afford" to be less cautious in the Star Wars Game, but—more to the point—their deterrent capability is undermined. Thereby, mutual preemption may become their jointly preferred equilibrium.

But observe that the exclusive MPE region requires that the defenses of the players be very strong—more than $\bar{c}$ on the horizontal axis and more than $\bar{r}$ on the vertical axis, as defined in the appendix and illustrated in figure 5.1. It is by no means clear that Star Wars technologies will ever permit such near-perfect defenses, so apprehensions about being in this exclusive MPE region are probably ill-founded, at least in the near future. Nevertheless, both the stability and the possible Pareto superiority of mutual preemption are alarming if Star Wars should ever approach perfection.

Figure 5.1 also shows that unilateral preemption by Column or Row give equilibria that overlap both DE and MPE. The payoff to the preemptor increases the weaker his opponent's first-strike defense, becoming 1 when his opponent has no Star Wars defense. Thus, each player has an incentive to be the sole preemptor, but only when his opponent's defense parameter, v, is less than $\bar{c}$ or $\bar{r}$; above these thresholds, MPE becomes the unique Nash equilibrium.

Although PE_C and PE_R cover much of the figure 5.1 square, we think they do not pose a serious threat now for two reasons. First, in the foreseeable future when Star Wars is likely to be far from perfect (well below c_3 and r_3 in figure 5.1), retaliatory threats will almost surely be a sufficient deterrent. Second, the PEs presume no retaliatory response on the part of the preempted player, but in fact the command and control procedures of the superpowers make a response almost automatic, as noted in earlier discussions of PDMs (sections 3.4, 3.5, and 4.2).

Superpower intentions reinforce their deterrent capabilities and virtually dictate the choice of DE over PE_C and PE_R in the regions of overlap. More worrisome is preemption—either unilateral or mutual —past the c_3 and r_3 points in figure 3.2. In fact, above these levels but below $\bar{c}$ and $\bar{r}$, when both sides' defenses can be considered moderately strong, each player prefers his PE to MPE, as shown in the appendix: Each side has the wherewithal to deter the other side and, in addition, does better when it preempts and threatens retaliation than when both sides preempt and neither threatens retaliation.

Beyond $\bar{c}$ and $\bar{r}$, when both sides' defenses are very strong, deterrence breaks down completely. Neither side can deter the other by threats. MPE is the unique Nash equilibrium and may even be Pareto-superior to (r_3,c_3), though certainly not when each side's Star Wars defenses are perfect so that the players' payoffs at MPE are exactly (r_3,c_3).

We believe it highly unlikely that the two superpowers will ever find themselves in either the exclusive MPE region or the region where the PEs and MPE overlap and both sides prefer a different PE (of course, Column always prefers PE_C to PE_R, and Row, PE_C to PE_R). Nevertheless, unilateral preemption is stable and always better for the preemptor, even in the DE region; this fact might conceivably threaten deterrence, especially in a crisis, if one side thinks the other is bluffing with its threats of retaliation. PDMs that support a retaliatory policy based on MAD or a similar doctrine, while frightening, probably remain the best means of preserving DE, at least in the early and middle stages of the development of a Star Wars system.

Scenario 2. *Second-strike defenses are perfect, first-strike defenses are variable:*

$$w_C = w_R = 1$$

When each side's defense against a second strike is perfect, as assumed in this scenario, DE is eliminated altogether. The reason, of course, is that neither side can credibly threaten a response if it has no capability to retaliate against a first strike.

Thus, only preemption equilibria show up in figure 5.2, with PE_C occurring when Row's Star Wars defense is relatively weak, and PE_R when Column's defense is relatively weak. MPE covers the rest of the area in figure 5.2, where both players' defenses are stronger.

There is no overlap between these equilibria, as there was in the previous scenario, because threats of retaliation are meaningless against a perfect second-strike defense. Without the ability to threaten, no PE can coexist with MPE; if one side preempts, it is better for the other side either to capitulate totally or to respond in kind.[7]

More precisely, if Row preempts when Column's level of defense is below c^*, or Column preempts when Row's defense is similarly weak, the preempted player is deterred from retaliation because he would do worse by moving the outcome toward the disastrous $(0,0)$ outcome in the Star Wars Game (as far as permitted by his opponent's defense). By comparison, above these thresholds deterrence fails, because the preempted player can do better by preempting also, which leads to MPE. It is shown in the appendix that the PEs and MPE in this scenario are all Pareto-superior.

Plainly, deterrence is undermined by a perfect second-strike defense, which would seem easier to approach than a perfect first-strike defense, the ultimate *raison d'être* of Star Wars. The principal reason is that after a first strike the ability of the attacked party to retaliate will be impaired, making it more difficult for this party to penetrate the attacker's second-strike defense rather than its first-strike defense.

Although an absolutely perfect Star Wars second-strike defense is probably an impossibility, it is useful to posit, nevertheless, in order to dramatize the extreme dependence of a deterrence equilibrium on the *lack* of a second-strike Star Wars defense. Certainly it is reasonable to assume that each side's first-strike defense will never be more effective than its second-strike defense, as in the present scenario.

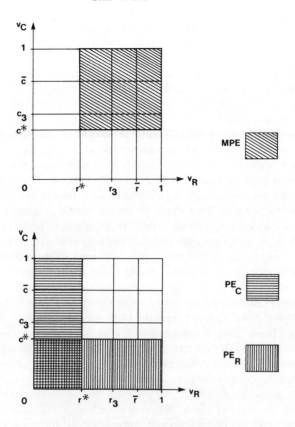

Key: v_C = level of Column's first-strike defense
v_R = level of Row's first-strike defense
MPE = mutual preemption equilibrium
PE_C = preemption equilibrium for Column
PE_R = preemption equilibrium for Row

Figure 5.2   Regions of existence of Nash equilibria in scenario 2: $w_R = w_C = 1$.

Furthermore, the factor of surprise lends additional plausibility to this assumption, for a retaliatory second strike can be better anticipated than a first strike, magnifying the effectiveness of a player's second-strike defense.

Scenario 3. *Second-strike defenses are never greater than first-strike defenses:*

$$w_C = v_C(1 - v_R); \qquad w_R = v_R(1 - v_C)$$

In one sense, this scenario is the opposite of scenario 2, wherein second-strike defenses were perfect and so could never be topped by first-strike defenses. Here second-strike defenses are never as good as first-strike defenses. Whereas scenario 2 wiped out the possibility of DE, our third scenario gives a substantial boost to DE, with scenario 1, in which first-strike and second-strike defenses are equal, an intermediate case.

Before offering a specific interpretation of the functional relationship between first-strike and second-strike defenses postulated in the present scenario, consider the circumstances under which first-strike defenses might actually exceed second-strike defenses. If space-based defenses are essentially invulnerable to a first strike, which they are meant to stymie, then these defenses would presumably be equally effective against a second strike, as assumed in scenario 1.

But now suppose that an attacker's first strike is largely counterforce. Then the attacked party's rational response will be largely countervalue, since it would be pointless for it to attack empty missile silos. Moreover, "soft" cities and industrial targets that are attacked in a countervalue strike are more vulnerable than "hard" missile sites and other protected defense targets that would be hit in a counterforce strike. Thus, Star Wars defenses against even a diminished second strike may actually be less effective than against a first strike, because the second strike (countervalue) is of a different character than the first strike (counterforce).

It is possible, therefore, to imagine that second-strike defenses may in fact be weaker than first-strike defenses, because strategic (and perhaps ethical) considerations may dictate different kinds of first and second strikes. One must be careful, however, in drawing any such comparison, because Star Wars defenses, as measured by the v's and w's, are based on different scales. First-strike defenses are anchored to a scale ranging from one's next-best to next-worst outcome, whereas second-strike defenses refer to a scale from one's best to worst outcome.

Our specific assumption in this scenario is that each player's second-strike defense is directly related to his first-strike defense (the better Star Wars is against first strikes, the better it is against second strikes) and inversely related to his opponent's first-strike defense (the more an opponent is able to stop a first strike, the harder he can hit back and penetrate one's second-strike defense). These assumptions are embodied in the simple functional relationship postulated above, which, we think, accords well with intuition, especially near the extremes when defenses are all or nothing.

For example, $w_C = 0$ (Column has no second-strike defense) if either $v_C = 0$ (Column has no first-strike defense) or $v_R = 1$ (Row has a perfect first-strike defense and hence can launch an unimpeded second strike). On the other hand, $w_C = 1$ (Column has a perfect second-strike defense) if both $v_C = 1$ (he has a perfect first-strike defense) and $v_R = 0$ (Row has no first-strike defense and hence will absorb the full force of Column's first strike).

Less defensible, perhaps, are the inverse relationships, which negatively link the second-strike defenses of one player to the first-strike defenses of the other. Admittedly, one's second-strike defenses are not *necessarily* helped as an opponent's first-strike defenses deteriorate, and vice versa. Yet the connection, even if indirect, is probably there: If an opponent cannot defend himself well when hit first, his retaliation will be degraded, thereby enhancing the attacker's second-strike defense. Basically, one's second-strike capability is incompatible with an opponent's first-strike capability: The greater the latter, the less the former in this scenario.

This third scenario is the most complex, and, not surprisingly, it exhibits the most subtle relationship among Nash equilibria. As shown in figure 5.3, DE, MPE, and both PE_C and PE_R overlap each other. In fact, all four can exist simultaneously, as when the Star Wars defenses of both players are at or above point (r^*, c^*) and more or less equal.

A symmetry in Star Wars defenses is good in this scenario, in part because DE always exists when the first-strike defenses of the players are reasonably close (i.e., near the $v_R = v_C$ diagonal in figure 5.3). Additionally, MPE never dominates DE in their region of overlap provided a certain condition, derived in the appendix, is met:

$$(1 - c_3)(1 - r_3) \leq (c_3 - c_2)(r_3 - r_2) \tag{5.1}$$

This condition may be interpreted to say that the *advantages* of unilateral preemption for the players, as embodied in the differences on the left side of (5.1), are less than the *damages* the players would suffer if preempted (on the right). A further consequence of (5.1) is that the region in which DE dominates MPE, located near the $v_R = v_C$ diagonal, extends all the way to $v_R = v_C = 1$.

We think condition (5.1) characterizes almost any conceivable nuclear conflict between the superpowers. Hence, it is virtually certain that DE will dominate MPE in this important region should the Star Wars defenses of the superpowers ever approach perfection.

As in scenario 1, if the defenses of the players are weaker (Row befow r^* and Column not too strong, or Column below c^* and Row not too strong), MPE does not exist when DE does. This salutary finding is counterbalanced, however, by the more ominous result that MPE, but not DE, can arise in the upper left and lower right parts of the rectangle with opposite corners at (r^*, c^*) and $(1,1)$, shown in the top part of figure 5.3.

In these regions, both players have moderately good Star Wars

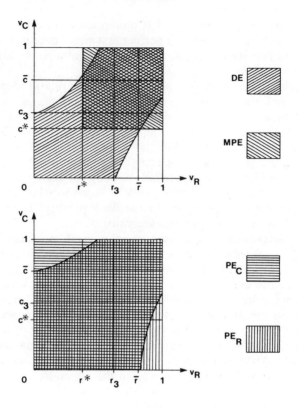

Key: v_C = level of Column's first-strike defense
      v_R = level of Row's first-strike defense
      MPE = mutual preemption equilibrium
      PE_C = preemption equilibrium for Column
      PE_R = preemption equilibrium for Row
      DE   = deterrence equilibrium

Figure 5.3   Regions of existence of Nash equilibria in scenario 3:
$$w_R = v_R(1-v_C); \quad w_C = v_C(1-v_R).$$

defenses, but one player's is significantly better than the other's. In such a situation, each player's best response to maximum preemption by his opponent is to preempt maximally himself, leading to MPE; neither player can possibly deter the other from preemption.

Slipping momentarily into one of these MPE regions may not be disastrous if the inferior player can quickly attain a better defense. Then the players' more symmetrical defenses would move them into a region that also admits DE, which would come to dominate MPE, subject to inequality (5.1). DE is able to reassert itself over MPE in this case by the equalization of the retaliatory capabilities of the players, with each being strong enough to deter the other when their v's (and therefore their w's) are more or less equal.

If DE is generally preferred to MPE where they overlap, what about PE_C and PE_R, which together completely cover the unit square and therefore always overlap with DE? Indeed, the two PEs overlap except when one player's Star Wars defense substantially exceeds the other's; then, as in the earlier scenarios, the stronger player can deter the weaker and especially benefit from unilateral preemption. Observe that as the inferior player catches up, however, the players move from exclusive PE regions to regions that admit both MPE and DE, but no PE is ever dominated by MPE or DE where they overlap.

Clearly, PE_C and PE_R are a terrible problem when one or both are the only Nash equilibria in a region. But it takes a comparatively large asymmetry in defense capabilities to trigger unilateral preemption, which seems most likely when one of the PEs is the unique equilibrium.

We suspect such an asymmetry is unlikely to persist once the inferior side realizes how precarious its position is, or might become, and takes steps to catch up. In this process of catching up, first both PE_C and PE_R appear (replacing just one), followed either by DE (at lower levels of defense), MPE (at higher levels), and both DE and MPE (at the highest levels). In the short term, DE will almost surely dominate MPE and, given PDMs, deflect players from the choice of unilateral preemption, but DE neither dominates nor is dominated by the PEs generally.

All three scenarios just described probably capture some part of the truth. In section 5.5 we compare them and assess the plausibility of their conclusions. We conclude by painting a composite picture of Star Wars and its potential strategic effects.

5.5 CONCLUSIONS

The Deterrence Game is founded on Chicken but allows players to choose any level of preemption at the start and threaten any level of retaliation if preempted. The Star Wars Game puts constraints on both the preemption (first-strike) and the retaliation (second-strike) levels that the players can choose in the Deterrence Game.

In the Deterrence Game, preemption by one player (PE_C and PE_R) and deterrence by both (DE) are the only Pareto-superior Nash equilibria. In the Star Wars Game, the constraints generally preserve the PEs and DE, but the regions in which they exist depend on the scenario postulated. An exception occurs in our second scenario, when both players have a perfect second-strike Star Wars defense. With retaliation against a first strike ruled out, deterrence becomes impossible and DE is wiped out.

In this scenario, mutual preemption (MPE) emerges as an equilibrium when the first-strike Star Wars defenses of both players exceed a particular threshold. At this equilibrium, because the players cannot be deterred by retaliatory threats, they do not make them. Instead, they find it better to attack simultaneously rather than absorb a first strike when their defenses are above certain threshold values.

We regard scenario 2 as the most unrealistic of the three, for perfect defenses, either first-strike or second-strike, simply do not seem in the cards. But scenario 2 does underscore the fragility of deterrence if Star Wars should ever approach perfection; in fact, the goal of SDI now seems to be to shore up deterrence rather than to provide a perfect defense (Mohr, 1986).

Yet deterrence could be undercut if an all-out first strike, with increasingly accurate nuclear weapons delivery systems, could cripple an opponent's retaliatory capacity. This possible negative relationship between one player's first-strike defense and his opponent's second-strike defense is explicitly modeled in our third scenario. Specifically, we assume that the stronger a player's first-strike defense (permitting him to hit back harder if attacked first), the weaker will be his opponent's second-strike defense (because he will have to suffer greater retaliation if he attacks first). This assumption, coupled with the assumption that each player's first-strike and second-strike defenses are positively related, leads to the substantial overlap among all equilibria in scenario 3.

Such overlap was anticipated in part from our first scenario, wherein we assumed strict equality between the players' first- and second-strike defense parameters. In this scenario, DE is Pareto-superior to MPE when the two equilibria overlap, but MPE ultimately displaces even the PEs as the Star Wars defenses approach leakproofness.

In scenario 1, DE vanishes precisely when both players' Star Wars defenses exceed the c_3 and r_3 thresholds necessary to deter second strikes, yielding to PEs and MPEs at higher levels of defense. Fortunately, these levels will probably be difficult to achieve, even in the distant future, so the disappearance of DE seems not to be a present danger if this scenario is an appropriate model.

The overlap of equilibria in scenario 3 preserved DE all the way up to both players' having perfect first-strike Star Wars defenses—as long as these defenses remain more or less equal (how near to equality depends on the players' payoffs). The most dangerous zones in this scenario occur when one player gains a substantial defensive lead over the other and when, at higher levels of defense, MPE may actually be Pareto-superior to DE. The latter possibility, however, seems remote, because near-perfect defenses are unlikely, and, even if they are attained, a necessary condition for MPE to dominate DE at this level is that a first strike produce unrealistically high rewards for the players.

All in all, our scenarios suggest that deterrence will remain viable as long as Star Wars defenses are so primitive that each side retains a substantial second-strike capability. As Star Wars becomes better and erodes these defenses, however, the primary danger lurks in one side's developing considerably stronger defensive capabilities than the other and thereby finding it rational to attack, secure that it can either deter retaliation completely or survive a ragged response and even be comparatively better off.

Preemption would be most advantageous for the superior player if he has a substantial defensive lead over the inferior player. Then, with DE eliminated, either PE (for the superior player) exists by itself or both PE and MPE exist.

Perhaps the greatest peril occurs when there is no deterrence equilibrium. Then a severe form of crisis instability may grip the players and lead them to an abyss. More probable in superpower relations, though, is that deterrence will remain reasonably secure, mainly because both sides have largely invulnerable second-strike capabilities (principally submarine-launched ballistic missiles and cruise missiles) that Star Wars will have no effect on, at least not presently.

Our major concern is that, short of being leakproof, Star Wars is probably more destabilizing than stabilizing. For one thing, it inevitably introduces MPE into the Star Wars Game, which did not exist in the Deterrence Game. For another, it shows, especially in scenario 3, that the unbalanced development of Star Wars capabilities by both sides is preemption-inducing and becomes more so as the Star Wars defense of the superior player approaches perfection. And, of course, Star Wars will be a horrendously costly venture, especially for such dubious returns.

If there is to be a full-fledged Star Wars Game, a time path for its development that keeps the two sides' defenses balanced is surely the best way to preserve deterrence.[8] Balance can probably most readily be achieved through the sharing of defensive technologies by the superpowers, as Ronald Reagan recommended in his 1983 speech on the subject and subsequently reiterated, even though others have ridiculed this idea as naive in the extreme and highly embarrassing.[9] Also, more frequent and open communication and better coordination of activities (e.g., military exercises) are obviously desirable to squelch fears of "breakout," whereby one side makes a major breakthrough against the other and thereby gets a big lead in the arms race.

At some point, however, perhaps in a severe crisis, crisis stability could be upset and preemption, perhaps even mutual preemption, might appear attractive. (We analyze such possibilities in chapter 7.) This has occurred at lower levels of conflict between the superpowers in different parts of the world. If we are to steer clear of *nuclear* preemption as a rational option, it is imperative that the superpowers recognize that they must carefully chart a course of balanced development—and, preferably, reach a verifiable agreement on no deployment—if Star Wars is not to jeopardize deterrence.

APPENDIX

The basic Star Wars Game is described in section 3.3. For future reference, we repeat the expected payoffs to Row and Column given in section 3.3:

$$E_R(s,q;t,p) = str_3 + (1-s)tp + s(1-t)qr_2 \tag{1}$$
$$E_C(t,p;s,q) = stc_3 + s(1-t)q + (1-s)tpc_2 \tag{2}$$

For the Star Wars Game, the strategic variables s, t, q, and p satisfy

$$v_C \leqq s \leqq 1, \qquad w_C \leqq q \leqq 1 \qquad \text{for Row}$$
$$v_R \leqq t \leqq 1, \qquad w_R \leqq p \leqq 1 \qquad \text{for Column}$$

Throughout this appendix, we will assume that the defense parameters satisfy

$$0 < v_R, \; v_C < 1; \qquad 0 < w_R, \; w_C \leqq 1$$

but our results do in fact apply when v_R and/or v_C is 0 or 1, or when w_R and/or w_C is 0. Our assumption is for convenience only, to avoid many tedious special cases.

The following lemma is easily vertified. It will prove useful in deriving Nash equilibria.

Lemma. Fix v and w such that $0 < v, w < 1$ and consider

$$f(x,y) = H + xK + xyL$$

where H, K, and L are constants with $L > 0$. Suppose that x and y must satisfy

$$v \leqq x \leqq 1; \qquad w \leqq y \leqq 1$$

Then

(i) $y = 1$ is necessary for a maximum of f.
(ii) $x = 1$, $y = 1$ maximizes f iff $K + L \geqq 0$.
(iii) $x = v$, $y = 1$ maximizes f iff $K + L \leqq 0$.
(iv) For any x' satisfying $v < x' < 1$, $(x,y) = (x',1)$ maximizes f iff $K + L = 0$.

The following definitions of r^* and $\bar{r}$ are based on Row's payoffs of r_2 and r_3. Analogous definitions of c^* and $\bar{c}$ will be assumed; of course, their properties are analogous as well.
Recall that

$$0 < r_2 < r_3 < 1$$

Set

$$r^* = \frac{r_2}{1 - r_3 + r_2}$$

It is easy to verify that

$$r_2 < r^* < r_3$$

Now let

$$Q(v) = v^2 - v(r_3 - r_2) - r_2$$

Because Q is quadratic with $Q(0) = -r_2 < 0$ and $Q(1) = 1 - r_3 > 0$, $Q(v)$ must have one zero in $(0,1)$. Denote this zero by $\bar{r}$.
Since $Q(r_3) = r_2(r_3 - 1) < 0$, it is clear that

$$0 < r_2 < r^* < r_3 < \bar{r} < 1$$

Furthermore, if $0 \leqslant v \leqslant 1$, then $Q(v) \leqslant 0$ iff $v \leqslant \bar{r}$, and $Q(v) = 0$ iff $v = \bar{r}$.

Deterrence Equilibrium

First we identify all Nash equilibria at which $s = 1$. If $s = 1$, (2) shows that

$$E_C(t,p;1,q) = q + t(c_3 - q) \tag{3}$$

so Column maximizes his expected payoff by choosing $t = 1$ iff $c_3 \geqslant q$. Since $q > w_C$, this is possible iff $w_C \leqslant c_3$. An analogous calculation for Row, and a straightforward verification, yields the deterrence equilibrium (DE):
I. There exists an equilibrium with $s = 1$ and $t = 1$ iff $w_C \leqslant c_3$ and $w_R \leqslant r_3$. In this case, all such equilibria are given by

$$s = 1, \, w_C \leqslant q \leqslant c_3; \quad t = 1, \, w_R \leqslant p \leqslant r_3 \tag{DE}$$

At any DE, the expected payoffs are $E_R = r_3$ and $E_C = c_3$.

Preemption Equilibria

Now observe from (4) and the fact that $t \geq v_R$ that if $s = 1$ and $q \geq c_3$, Column maximizes his expected payoff by choosing $t = v_R$. By (1),

$$E_R(s,q;v_R,p) = v_R p + s[v_R r_3 - v_R p] + sq[(1 - v_R)r_2]$$

The Lemma now shows (i) that $q = 1$ is necessary to maximize E_R and (ii) that $s = 1$, $q = 1$ maximizes Row's expected payoff iff

$$v_R r_3 - v_R p + (1 - v_R)r_2 \geq 0$$

which is equivalent to

$$p \leq \frac{v_R r_3 + (1 - v_R)r_2}{v_R}$$

Since $w_R \geq p$, this means that

$$w_R \leq \frac{v_R r_3 + (1 - v_R)r_2}{v_R} \tag{4}$$

Inequality (4) is not an effective restriction on w_R if the right side is at least 1; it is easy to verify that this occurs iff $v_R \leq r^*$. This leads to Column's preemption equilibrium (PE_C):

II. There exists an equilibrium with $s = 1$ and $t = v_R$ iff either $v_R \leq r^*$ or $v_R > r^*$ and

$$w_R \leq \frac{v_R r_3 + (1 - v_R)r_2}{v_R}$$

In this case, all such equilibria are given by

$$s = 1, \ q = 1; \qquad t = v_R, \ w_R \leq p \leq \frac{v_R r_3 + (1 - v_R)r_2}{v_R} \tag{PE_C}$$

At any PE_C, the expected payoffs are $E_R = v_R r_3 + (1 - v_R)r_2$ and $E_C = 1 - v_R(1 - c_3)$.

It is instructive to consider the implications of II in the case of the functional relationship in scenario 3, $w_R = v_R(1 - v_C)$. Substitution in (4) and simplification leads to the condition

$$v_C \geq \frac{Q(v_R)}{v_R^2}$$

where $Q(v_R)$ is as defined above. It follows that if $w_R = v_R(1 - v_C)$, PE_C exists either if $v_R \leq \bar{r}$, or if $v_R > \bar{r}$ and $v_C \geq Q(v_R)/v_R^2$.

We now show that I and II define the only equilibria with $s = 1$. For an equilibrium with $v_R < t < 1$, (2) implies that

$$\frac{\partial E_C}{\partial t}(t,p\,;1,q) = c_3 - q = 0$$

is necessary, so $q = c_3$ at an equilibrium with $s = 1$ and $v_R < t < 1$. But (1) yields

$$\frac{\partial E_R}{\partial q}(s,q\,;t,p) = s(1 - t)r_2 > 0$$

since $t < 1$, so $q = 1$ at equilibrium. This contradiction shows that there are no equilibria with $s = 1$ other than I and II.

By symmetry, we conclude that the only equilibria with $t = 1$ are DE and PE_R; the latter exists iff either $v_C \leq c^*$ or $v_C > c^*$ and

$$w_C \leq \frac{v_C c_3 + (1 - v_C)c_2}{v_C}$$

It consists of all strategy combinations

$$s = v_C, \quad w_C \leq q \leq \frac{v_C c_3 + (1 - v_C)c_2}{v_C}; \qquad t = 1, p = 1 \qquad (PE_R)$$

At any PE_R, the expected payoffs are $E_R = 1 - v_C(1 - r_3)$ and $E_C = v_C c_3 + (1 - v_C)c_2$.

Mutual Preemption Equilibrium

We now search for equilibria with $s = v_C$ and $t = v_R$. By (1),

$$E_R(s,q\,;v_R,p) = v_R p + s[v_R r_3 - v_R p] + sq[(1 - v_R)r_2]$$

The Lemma shows that (i) $q = 1$ is necessary to maximize E_R and (iii) $s = v_C$, $q = 1$ maximizes E_R iff

$$v_R r_3 - v_R p + (1 - v_R)r_2 \leq 0$$

This is equivalent to

$$p \geq \frac{v_R r_3 + (1 - v_R)r_2}{v_R}$$

Because $p \leqq 1$, a necessary condition for equilibrium is

$$v_R r_3 + (1 - v_R) r_2 \leqq v_R$$

which is easily seen to be equivalent to $v_R \geqq r^*$.

An analogous calculation for Column's expected payoff, and a simple verification, give the mutual preemption equilibrium (MPE): III. There is an equilibrium with $s = v_C$ and $t = v_R$ iff $v_C \geqq c^*$ and $v_R \geqq r^*$. In this case, the only equilibrium is

$$s = v_C, q = 1; \qquad t = v_R, p = 1 \tag{MPE}$$

At any MPE, the expected payoffs are

$$E_R = v_R v_C r_3 + v_R(1 - v_C) + (1 - v_R) v_C r_2$$

and

$$E_C = v_R v_C c_3 + (1 - v_R) v_C + v_R(1 - v_C) c_2$$

Line Equilibria

We now search for equilibria with $s = v_C$ and $v_R < t < 1$. By (2),

$$E_C(t, p; v_C, q) = v_C q + t[v_C c_3 - v_C q] + tp[(1 - v_C) c_2]$$

The Lemma shows that (i) $p = 1$ at any equilibrium and (iv) $v_R < t < 1$ at an equilibrium iff

$$v_C c_3 - v_C q + (1 - v_C) c_3 = 0$$

which is equivalent to

$$q = \frac{v_C c_3 + (1 - v_C) c_2}{v_C} \tag{5}$$

But

$$\frac{\partial E_R}{\partial q}(s, q; t, p) = s(1 - t) r_2$$

which implies that $q = 1$ at any equilibrium with $t < 1$. Combining with (5) gives the necessary condition,

$$v_C = \frac{c_2}{1 - c_3 + c_2} = c^*$$

Applying the Lemma to E_R under the conditions $v_C = c^*$, $v_R < t < 1$, and $p = 1$ shows that $t \geqq r^*$ is necessary at any equilibrium with $s = v_C$. It is straightforward to verify Row's line equilibrium (LE_R):

IV. There is an equilibrium with $s = v_C$ and $v_R < t < 1$ iff $v_C = c^*$. In this case, all such equilibria are given by

$$s = v_C = c^*, q = 1; \qquad v_R < t < 1, t \geqq r^*, p = 1 \qquad (LE_R)$$

At any LE_R, the expected payoffs are

$$E_R = t - (t - r^*)(1 - r_3 + r_2)c^* \qquad \text{and} \qquad E_C = c^*$$

Between two different LE_R equilibria, Row prefers the one with the larger value of t, and Column is indifferent.

As before, we note that there is an equilibrium LE_C analogous to LE_R, with $t = v_R$ and $v_C < s < 1$ iff $v_R = r^*$. All such equilibria are then given by

$$v_C < s < 1, s \geqq c^*, q = 1; \qquad t = v_R = r^*, p = 1 \qquad (LE_C)$$

The expected payoffs at LE_C are analogous to those at LE_R.

Naive Equilibrium

Finally, we search for equilibria with $v_C < s < 1$ and $v_R < t < 1$. By differentiating (2), it is easy to see that $p = 1$ at such an equilibrium, giving

$$E_R(s, q; t, 1) = t + s[tr_3 - t] + sq[(1 - t)r_2]$$

The Lemma shows (iv) that Row's best response to $v_R < t < 1$ and $p = 1$ is $v_C < s < 1$ and $q = 1$ iff

$$tr_3 - t + (1 - t)r_2 = 0$$

which is equivalent to $t = r^*$. It follows from $v_R \leqq t$ that $v_R \leqq r^*$ is necessary for this equilibrium. A similar calculation for Column's expected payoff, and a simple verification, yield the naive equilibrium (NE):

V. There is an equilibrium with $v_C < s < 1$ and $v_R < t < 1$ iff $v_R \leqq r^*$ and $v_C \leqq c^*$. In this case, the unique such equilibrium is

$$s = c^*, q = 1; \qquad t = r^*, p = 1 \qquad (NE)$$

The expected payoffs are $E_R = r^*$ and $E_C = c^*$.

Based on the definitions of equilibria I–III, the drawing of figures 5.1 and 5.2 for scenarios 1 and 2 in section 5.4 is straightforward. In the case of scenario 3, where $w_C = v_C(1 - v_R)$ and $w_R = v_R(1 - v_C)$, the construction of the curves in figure 5.3 may require some explanation. As a consequence of I, the region of existence of DE in figure 5.3 is bounded by the curves $v_C = c_3/(1 - v_R)$ (upper left) and $v_R = r_3/(1 - v_C)$ (lower right). As indicated above, the boundary of the PE_R region in figure 5.3 is $v_C = Q(v_R)/v_R^2$, and the boundary of the PE_C region is $v_R = Q(v_C)/v_C^2$.

Comparisons of Payoffs at Different Equilibria

We now turn to the question of the Pareto superiority of the various Nash equilibria. It is easy to show that Row always prefers PE_R to DE to PE_C, and Column always prefers PE_C to DE to PE_R, so none of these three equilibria ever dominates any other. It can readily be verified that DE is better for both players than NE and that all line equilibria are always dominated by the preemption equilibrium that is their limiting case. (For example, LE_R is dominated by PE_C.) Thus the Pareto-superior equilibria are always to be found among DE, PE_R, PE_C, and MPE.

The status of MPE is especially interesting. Straightforward calculations give

$$E_R(PE_R) - E_R(MPE) = (1 - v_R) \, [1 - v_C(1 - r_3 + r_2)] > 0$$
$$E_C(MPE) - E_C(PE_R) = (1 - v_R) \, (1 - c_3 + c_2) \, (v_C - c^*) \geqq 0$$

since MPE does not exist unless $v_C \geqq c^*$. Similar computations for PE_C yield that none of PE_R, PE_C, and MPE is Pareto-inferior to any of the others.

The comparison of MPE and DE is more complex. It is easy to see that

$$E_R(MPE) > E_R(DE) \qquad \text{iff} \qquad v_R > f_C(v_C) = \frac{r_3 - v_C r_2}{1 - v_C(1 - r_3 + r_2)}$$

Now $f_C(v_C)$ is a smoothly increasing function satisfying $f_C(0) = r_3$ and $f_C(1) = 1$. It follows that $E_R(DE) \geqq E_R(MPE)$ if $v_R \leqq r_3$; Row prefers MPE to DE only if v_R is large enough. Analogous conclusions can be drawn for Column's payoffs.

It follows that MPE dominates DE exactly when $v_R > f_C(v_C)$ and $v_C > f_R(v_R)$ hold simultaneously, and DE dominates MPE when both inequalities fail simultaneously. Consideration of the derivatives $f_C'(1)$ and $f_R'(1)$ shows that MPE dominates DE iff the two equilibria coexist in a region near $(v_R, v_C) = (1,1)$, and $(1 - r_3)(1 - c_3) > (r_3 - r_2)(c_3 - c_2)$.

The region in which DE dominates MPE includes $v_R \leq r_3$, $v_C \leq c_3$, as noted above. It intersects every neighborhood of $(v_R, v_C) = (1,1)$ iff

$$(1 - r_3)(1 - c_3) \geq (r_3 - r_2)(c_3 - c_2)$$

NOTES

This chapter is drawn from Steven J. Brams and D. Marc Kilgour, Deterrence versus defense: a game-theoretic model of Star Wars, *International Studies Quarterly* 32, no. 1 (March 1988). Reprinted with permission of the International Studies Association, Byrnes International Center, University of South Carolina, Columbia, SC 29208 USA.

1 A sophisticated discussion of the issues SDI has raised, and a careful weighing of the arguments of both proponents and opponents, can be found in Office of Technology Assessment (1985). Stability questions about SDI particularly relevant to this chapter are discussed in Marsh (1985). More general discussions of SDI and its potential effects can be found in a number of recent books, including Wells and Litwak (1987); Bowman (1986); Payne (1986); Brzezinski (1986); Stever and Pagels (1986); Drell, Farley, and Holloway (1985); Thompson (1985); Miller and Van Evera (1986); Dallmeyer (1986); and *Weapons in Space* (1985).

2 See, for example, Chrzanowski (1985a, 1985b); Kerby (1986); Canavan (1985); Wilkening and Watman (1986); Max et al. (1986); Saperstein and Mayer-Kress (1987); DeNardo (1987); and Bracken (1986). Bracken shows, among other things, how the one-sided deployment of SDI can create an incentive for the side deploying to strike first; his analysis anticipates some of our conclusions, including the problem of avoiding preemption in the transition to full deployment. While Bracken's model is not game-theoretic, O'Neill (1987) defines axiomatically a game-theoretic index of crisis instability and applies it to SDI.

3 This upper bound is not an effective restriction if $v_R \leq r^*$, where $r^* = r_2/(1 - r_3 + r_2)$, because then it is at least 1. In specifying the upper bound, we assume that it is not less than the lower bound (so that there exists a p that satisfies the double inequality for p given in the definition of PE_C) and that $v_R > r^*$. (It can be shown that $r_2 < r^* < r_3$.) Details are given in the appendix.

4 Note that when $v_R = 0$ so that Row has no first-strike Star Wars defense, PE_C in the Star Wars Game is PE_C in the Deterrence Game.

5 Presumably, if one player has arms that are totally ineffective against an opponent, he will not use them. Instead, it is reasonable to suppose that he will save them for some other use, perhaps against another opponent.

6 In the appendix we show that neither PE nor MPE may dominate the other. If they coexist, DE and MPE may both be Pareto-superior, or either may dominate the other. In any case, there is always at least one Pareto-superior equilibrium, and all Pareto-superior equilibria are either DE, PE_R, PE_C, or MPE.

7   In the first scenario, "response in kind" could sometimes be deterred by threats. The possibility of retaliation permits PEs to coexist with MPE, which never involves threats.

8   Nevertheless, Glaser (1987) contends that the transition period will present formidable difficulties. Commenting on Glaser, Carnesale (1987, p. 175) captures well the issue we have tried to model here: "If both side are going to have ballistic missile defense, then the Soviets may have increased uncertainty about their first strike and we would no doubt have increased uncertainty about our ability to retaliate. Would we be better off or worse off?"

9   Ignatius (1985); Sanger (1985); Gwertzman (1985); and Scowcroft, Deutsch, and Woosey (1987). The sharing of technology for "permissive action links" (PALs), or coded locks for nuclear weapons, has, by contrast, been universally applauded because "everyone benefits if nuclear weapons are kept under better control" (Marshall, 1986).

6

Optimal Threats

Probably the best-known precept of justice in the Bible is "eye for eye, tooth for tooth, hand for hand, foot for foot" (Exod. 21:24), which is applied also to homicide:

> Whoever sheds the blood of man,
> By man shall his blood be shed. (Gen. 9:6)

Although it is commonly abbreviated "an eye for an eye, a tooth for a tooth," we henceforth refer to this precept as *tit-for-tat*.

Much harsher retribution is called for elsewhere in the Bible by a God who "does not remit all punishment but visits the iniquity of fathers upon children and children's children, upon the third and fourth generation" (Exod. 34:7). Yet the golden rule eschews retribution: "You shall not take vengeance or bear a grudge against your kinsfolk. Love your neighbor as yourself" (Lev. 19:18). To be sure, the golden rule suggests the same kind of reciprocity as tit-for-tat, but emphatically without the threat of punishment; charity is the watchword here. Altogether, this bewildering variety of prescriptions —taken from just three books of the Hebrew Bible, all part of the Torah or Pentateuch[1]—indicates that even in biblical times there was little consensus on what constituted fair treatment or just and reasonable punishment.

There has been renewed interest in tit-for-tat since Axelrod's *The Evolution of Cooperation*, which reported results of two experiments in which computer programs specifying strategies were matched against each other in tournament play of Prisoners' Dilemma.[2] Axelrod's conclusions concerning the robustness of tit-for-tat, it should be emphasized, are in large part empirical and follow from a model that assumes repeated play against different (randomly chosen) opponents. By comparison, we propose an analytic model in which aggressive actions are immediately known to the players (that is, not inferred from repeated play), there may be varying levels of aggression and retaliation, and there are two rather than many different players, as in a tournament. Moreover, the model we develop is not based on Prisoners' Dilemma (although it is applicable to this game; see note 5) but instead on Chicken, the game underlying the Deterrence and Winding-Down Games analyzed in chapters 3 and 4.[3]

Our main purpose in this chapter is not to prescribe appropriate or fair punishment against aggression. Rather, it is to determine what threatened level of punishment is optimal in *deterring* aggression. In contradistinction, punishing aggression after it has occurred signifies a failure of deterrence. If deterrence fails and the threatened retaliation occurs, we assume that *both* players will be hurt. Our calculation of optimal threats to deter aggression will focus on those threats that do so at minimal cost to the threatener but still make it rational for the threatened player *not* to attack the threatener in the first place.

The principal question we address here is under what conditions the threatened level of retaliation should be less than proportionate, proportionate (i.e., tit-for-tat), or more than proportionate to deter aggression in two-person conflicts that can be modeled by a game based on Chicken. As we will show, the answer might be any of these, depending on the parameters of Chicken and assumptions about threats of retaliation and expected payoffs that we introduce into the basic structure of this game.

Perhaps our most important result is that a policy of tit-for-tat may be too crude. Indeed, if one's purpose is to deter aggression at potentially minimum cost, one must tailor the threatened punishment to the level of provocation or aggression, but the relationship is not a simple one: Sometimes the threatened retaliation should be more than proportionate, sometimes exactly proportionate, and sometimes less than proportionate.

In general, as the level of provocation rises, retaliation should increase, but at a decreasing rate. Thus, if the level of threatened retaliation starts off as more than proportionate, eventually, as the provocation increases, a point will be reached beyond which less-than-proportionate retaliation will be sufficient to deter the potential aggressor.

We begin our basic analysis by describing the Threat Game, which is a cousin of the Deterrence Game. The Threat Game is also based on Chicken but differs from the Deterrence Game in a crucial way. We then characterize Nash equilibria in this game and discuss optimal strategies associated with deterring an opponent. The relevance of these strategies to historical cases and contemporary international relations, particularly to maintaining nuclear deterrence between the superpowers, will be considered. Finally, we shall offer some brief remarks on a normative theory of deterrence based on the revisions suggested in tit-for-tat.

6.2 THE THREAT GAME

We argued in chapter 3 that Chicken, with the possibility of threats of retaliation incorporated into its structure, constitutes an appropriate game for analyzing deterrence, particularly nuclear deterence between the superpowers. The Threat Game adds two significant features to Chicken that make it an excellent model for investigating the nature of threats necessary to sustain deterrence: (1) Like the Deterrence Game, the players can make quantitative choices of levels of cooperation (C) or noncooperation ($\bar{C}$), not just qualitative choices of C or $\bar{C}$; (2) unlike the Deterrence Game, once these initial choices, which can be interpreted as levels of nonpreemption (versus preemption) are made, only the less preemptive player (i.e., the player who chose the lower level of preemption initially, if there was one) can retaliate by choosing a different, and presumably higher, level of noncooperation subsequently. The game terminates whenever the initial levels are the same or, if they are not, after the less preemptive player has retaliated.

To be sure, the rule that allows only the more aggrieved player, who chose a lower level of preemption initially, to retaliate is arbitrary. Our reasoning is that conflict escalates precisely because each player, in turn, is willing and able to raise the ante. If escalation were unilateral and one player simply became more and more aggressive without provoking retaliation from the other, then the process would not be two-sided escalation but one-sided aggression.

In this chapter, we are interested in modeling situations in which both players move up the escalation ladder. However, we telescope this process into one in which, after initial preemptive moves by the two players, a single retaliatory countermove is allowed that can "right the balance" for the initially more cooperative player. (Other rounds of escalation could be added, but we think the present simple sequence captures well both the process that might trigger further escalation and the core meaning of deterrence: averting conflict through the threat of retaliation, which, if carried out, could be costly to both players.)

Recall that in the Deterrence Game we did not permit retaliation to be a function of the level of preemption of the more preemptive player. Instead, we showed that deterrence could be achieved by threats of retaliation that did not depend on the actual level of preemption. The introduction here of *variable* retaliation in the Threat Game may be viewed as a refinement of this model in that it ties retaliatory threats to the level of preemption rather than presuming that only one threat—intended to deter any level of preemption—applies to all situations. In a real-life context, this refinement is echoed in the abandonment by the United States of "massive retalition" in favor of "flexible response" and "graduated deterrence" around 1960 precisely because the former doctrine was deemed too blunt (and incredible) a retaliatory policy for anything except an all-out Soviet nuclear attack.

More formally, the Threat Game is defined by the following rules:

1 The players do not choose initially between C and $\bar{C}$ but instead choose nonpreemption levels s (Row) and t (Column) that may range on a continuum between 1 (no preemption) and 0 (maximum preemption). These choices are simultaneous and determine a point on the unit square defined by vertical coordinate $y = s$ and horizontal coordinate $x = t$ (see figure 6.1), which we call the *initial position*. It is convenient to assume that before play commences the *preplay position* of the players is (1,1), the mutual-cooperation position, but this starting point is not crucial to the analysis of the Threat Game. In fact, the route by which the "final position" (to be defined in rule 4) is attained is irrelevant to the payoffs the players can achieve.

2 The 2×2 game of Chicken is embedded in the unit square so that the four corners of the square correspond to the four "pure" states (r_i, c_j) of Chicken, with r_i giving the payoff to Row and c_j the payoff to Column. The subscripts i and j give the rankings of payoffs, with 4 being best and 1 being worst; these rankings are given in the figure 3.1 representation of Chicken. Payoffs to the players are assumed to be normalized, so that the best and worst payoffs are 1 and 0, respectively:

$$0 = r_1 < r_2 < r_3 < r_4 = 1; \qquad 0 = c_1 < c_2 < c_3 < c_4 = 1$$

3 In figure 6.1, the square is drawn so that the (1,1) position corresponding to the CC outcome of Chicken is at the upper left and the (0,0) position corresponding to the $\bar{C}\bar{C}$ outcome of Chicken is at the lower right. (This positioning of outcomes corresponds to the usual one for 2×2 games, but it reverses the usual orientation of the horizontal axis: Values of x increase from right to left.) Points of the square not at the four corners represent mixtures of two or more outcomes of Chicken; the payoffs that these mixtures yield are defined in rule 5 and illustrated subsequently.

4 If $s = t$, the game terminates at point (t,s) in the unit square. If $s > t$, the more cooperative player, Row, retaliates against Column by changing his level of cooperation, according to the retaliation function $q(t)$, to make the *final position* of the game $(t, q(t))$. Row's retaliation is thus a function of Column's initial choice t; however, it is the value of s, as compared with t, that determines whether Row or Column (if either) can retaliate. If $s < t$, Column is the player who is more cooperative initially and thereby can retaliate; Column's level of cooperation changes from t to $p(s)$,

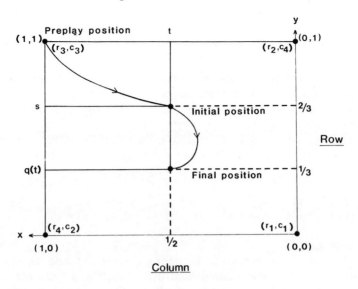

Figure 6.1 Threat Game.

Key: (r_i, c_j) = (payoff to Row, payoff to Column)

r_4, c_4 = best; r_3, c_3 = next best; r_2, c_2 = next worst; r_1, c_1 = worst

s, t = initial strategy choices of Row and Column, respectively

$q(t)$ = subsequent strategy choice of Row (more cooperative player initially)

Normalization: $0 = r_1 < r_2 < r_3 < r_4 = 1$; $0 = c_1 < c_2 < c_3 < c_4 = 1$

where $p(s)$ is Column's retaliation function, and the final position is $(p(s), s)$. The players' retaliation functions are chosen at the same time as their initial choices, s and t. (Much of the rest of this chapter is devoted to showing what initial choices and retaliation functions are, in a certain sense, optimal for the players in the Threat Game.)

5 Payoffs to Row (R) and Column (C) are defined by both the Chicken payoffs at the four corners of the unit square and the final position (x, y) of the players on the square:

$$P_R(x,y) = xyr_3 + x(1-y)r_4 + (1-x)yr_2 + (1-x)(1-y)r_1$$
$$= x + yr_2 - xy(1 - r_3 + r_2) \quad \text{(after normalization)}$$
$$P_C(x,y) = y + xc_2 - xy(1 - c_3 + c_2)$$

At the four corners of the unit square, P_R and P_C coincide exactly with the corresponding payoffs of Chicken. At any other point on the square, P_R and P_C are averages of the four payoffs at the corners, weighted according to the distances, measured horizontally and vertically, of the point from the corner.

It is worth noting that the payoff functions are bilinear (linear in each coordinate): In calculating payoffs at (x,y), the payoffs at each corner are weighted by the product of the distances, parallel to the axes, from (x,y) to the opposite corner.[4] As an illustration of this weighting system, consider the coordinates of the final position shown in figure 6.1, where $x = 1/2$ and $y = 1/3$. The weights of the four corner values of Chicken in $P_R(x,y)$ and $P_C(x,y)$ are

$(1,1)$: $xy = 1/6$ $(0,1)$: $(1-x)y = 1/6$
$(1,0)$: $x(1-y) = 1/3$ $(0,0)$: $(1-x)(1-y) = 1/3$

which sum to 1.

The two lower payoffs in figure 6.1, (r_4,c_2) and (r_1,c_1), which correspond to corners equidistant from $(1/2,1/3)$, are equally weighted by factors of $1/3$ in P_R and P_C. Similarly, the two upper payoffs, (r_3,c_3) and (r_2,c_4), which also correspond to corners equidistant from $(1/2,1/3)$, are equally weighted by factors of $1/6$. Not unreasonably, the upper factors have half as much "pull" as the lower factors, for they are twice as distant from $(1/2,1/3)$ on the vertical axis.

Different weighting factors would deform the continuous surface generated by our payoff functions, but these deformations would remain anchored at the four corners of the unit square. They would not alter the basic nature of the Nash equilibrium results that we describe in section 6.3.

An advantage offered by the present formulation, in addition to mathematical tractability, is that the four weights in P_R and P_C can be interpreted as probabilities because they are non-negative and sum to 1. Consequently, P_R and P_C may be thought of as expected values, the corner values being the four "pure" states of the game that can arise. As probabilities of being at the four corners, the four factor weights, when multiplied by a player's payoffs at each corner, give his expected payoff at any point of the unit square. It is worth noting that these factor weights are functions of the mixed (i.e., probabilistic) strategies of the players, who can be viewed as choosing strategy mixtures $(x,1-x)$ and $(y,1-y)$.

In a nonprobabilistic interpretation, which we prefer because it leaves unambiguous who initially is the more cooperative player, the unit square of figure 6.1 might be considered a game board. The players choose x,y coordinates, according to rules 1–4, that define a point on this board. As given by rule 5, the payoffs to the players are the heights of the continuous surfaces, above the xy plane, generated by P_R and P_C.

6.3 NASH EQUILIBRIA IN THE THREAT GAME

A Nash equilibrium in the Threat Game has the property that the player who was more cooperative initially cannot increase his payoff (at the final outcome) by changing his coordinate when his opponent's coordinate is held constant. When the players are equally cooperative (or aggressive) initially, ruling out retaliation on the part of either, neither can do better by increasing or decreasing his level of preemption when his opponent's is held constant.

The Threat Game includes several qualitatively different kinds of Nash equilibria, but only three classes are listed below. This list includes all undominated equilibria; dominated equilibria—for which there is another equilibrium at which both players do at least as well and at least one player does better—are excluded. The undominated equilibria may be thought of as those that are Pareto-superior within the set of all equilibria.

For the sake of completeness, we include in the appendix a list of *all* Nash equilibria of the Threat Game. Only undominated equilibria, however, are formally derived in the appendix.

For each of the three classes of undominated equilibria, the x,y coordinates of the equilibrium final position are given, then the payoffs to the players, and finally the strategies that led to the equilibrium. Brief verbal descriptions are also included. The classes are organized according to the final position of the equilibrium.

I *Deterrence equilibrium*: $(x,y) = (1,1)$, with payoffs (r_3,c_3) to Row and Column.

The initial strategies and retaliation functions that define equilibria of this class are

$$s=1, \ q(t) \leq q_1(t); \qquad t=1, \ p(s) \leq p_1(s)$$

where

$$q_1(t) = \frac{c_3 - tc_2}{1 - t(1 - c_3 + c_2)}, \qquad 0 \leq t < 1$$

and

$$p_1(s) = \frac{r_3 - sr_2}{1 - s(1 - r_3 + r_2)}, \qquad 0 \leq s < 1$$

At the deterrence equilibrium (DE), both players never preempt $(s = t = 1)$, but Row threatens retaliation if preempted at any level $t < 1$, and Column threatens similarly if $s < 1$.

If, say, Column were to preempt Row by choosing his coordinate $x = t < 1$, Row would respond by changing his coordinate from $y = s = 1$ to $y = q(t) \leq q_1(t)$. The 1 subscripts on the right-hand sides of each inequality indicate that the preempted player initially chooses $s = 1$ (Row) or $t = 1$ (Column). Related retaliation functions are discussed under (III) Line equilibria. Note that the retaliation levels (q and p) that Row and Column threaten are functions of the opponent's level of preemption (t and s). We will have more to say about these levels in section 6.4.

II *Preemption equilibria*:

(1) $(x,y) = (1,0)$, with payoffs $(1,c_2)$

(2) $(x,y) = (0,1)$, with payoffs $(r_2,1)$

Call (1) the preemption equilibrium of Row (PE_R); it is defined by $s = 0$, $t > 0$, $p(s)$ arbitrary, except $p(0) = 1$. Analogous conditions define (2), the preemption equilibrium of Column (PE_C). These equilibria are shown in figure 6.2.

PE_R says that Row preempts at his maximum level ($s = 0$). Column may preempt at any level $t > 0$, and his retaliation function $p(s)$ is arbitrary for $s > 0$, but $p(0) = 1$ is required: Column responds to Row's maximum preemption by remaining at $t = 1$ if he started out there or, if $0 < t < 1$, retreating totally from his initial preemption level to full cooperation (at $x = 1$). Note that PE_R requires $t > 0$, for if t were 0 when $s = 0$, the game would end at $(x,y) = (0,0)$ instead of $(1,0)$ since neither player can change his initial choice if $s = t$.

III *Line equilibria*:

(1) $(x,y) = (1,y)$, $0 < y \leq y^* = c_2/(1 - c_3 + c_2)$,
     with payoffs $(1 - y + r_3 y, c_2 - c_2 y + c_3 y)$

(2) $(x,y) = (x,1)$, $0 < x \leq x^* = r_2/(1 - r_3 + r_2)$,
     with payoffs $(r_2 - r_2 x + r_3 x, 1 - x + c_3 x)$

Call (1) the line equilibria for Row (LE_R); for each y satisfying $0 < y \leq y^*$, LE_R is defined by

$$s = y, \ q(t) \text{ arbitrary}; \qquad t = 1 \text{ or } y < t \leq 1 - y + y r_3, \ p(s) \text{ arbitrary}$$

except $p(y) = 1$ and $p(s) \leq p_y(s)$ for $0 \leq s < y$, where

$$p_y(s) = \frac{1 - y + y r_3 - s r_2}{1 - s(1 - r_3 + r_2)}, \qquad 0 \leq s < y$$

LE_R says that Row partially preempts at level $s = y < 1$, and Column either does not preempt ($t = 1$) or is less preemptive than Row such that $y < t \leq 1 - y + y r_3$. By the rules of the Threat Game, therefore, Column has the right to respond. In the actual event, Column's response to $s = y$ is to remain at, or retreat to, $x = 1$.

If Row were to choose $1 > s > y^*$, where y^* is defined above (see

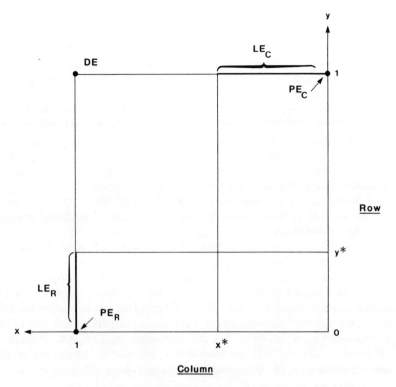

Note: The final positions of only undominated equilibria are shown.

Figure 6.2   Equilibrium final positions in Threat Game.

also figure 6.2), it would always be rational for Column to retaliate in a manner that hurts Row. Because Row would therefore have an incentive to choose a different initial strategy to avoid or deter retaliation, there are no equilibria along the line $x = 1$ above $y = y^*$ in figure 6.2.

On the other hand, if $0 < s \leq y^*$, Column does better by *not* retaliating. Hence, for any choice of $s = y$ in this interval, the retaliation function of Column that is in equilibrium has $p(y) = 1$ (i.e., no retaliation).

In effect, the line equilibria are "extensions" of the two preemption equilibria; as $y \to 0$, the line equilibria for Row become Row's preemption equilibrium, and similarly for Column. Until the limit is reached, however, the LEs are outcomes of less-than-complete preemption stabilized by an opponent's retaliatory threats against still greater levels of preemption.

In the case of the deterrence equilibrium, the choice of $p_1(s)$ by

Column or $q_1(t)$ by Row associated with DE is sufficient to make *any* level of preemption by the opponent as costly as no preemption, so the opponent cannot do better than choose his own DE strategy. Since one player can, by his choice of his DE strategy, make the compromise outcome uniquely attractive for his opponent, and vice versa, there is a strong reason for both players to settle on DE—rather than on any of the other Nash equilibria—unless the capabilities of the players are very different (in which case our model, based symmetrically on Chicken, would not be applicable). Like the deterrence equilibrium in the Deterrence Game, however, DE in the Threat Game is imperfect: If deterrence for any reason should fail, it may be irrational to retaliate—for Column, if $s = y < y^*$, as discussed earlier—since retaliation would imply a worse outcome for the threatener, having to carry out his threat, as well as for the player who preempted and thereby provoked retaliation.

As in the Deterrence Game, the problem of imperfectness is "solved," in a sense, if the players can irrevocably precommit themselves to carrying out their threats. For the superpowers, in fact, this seems indeed to be the case, as we argued in section 3.4. In modeling their conflict by the Threat Game, therefore, we assume that they can make precommitments that are regarded as credible: The threatened player believes that the threatener will in fact retaliate at the level he says he will. In section 6.4, we take a closer look at DE and the nature of the retaliatory threats necesssary to sustain it.

6.4 OPTIMAL STRATEGIES IN THE THREAT GAME

DE is an equilibrium with final position $(x,y) = (1,1)$. At this equilibrium, each player is deterred from reducing his initial level of cooperation below $s = t = 1$ by his opponent's threat of retaliation. For example, because $p(s) \leq p_1(s)$, Row does at least as well by choosing $s = 1$ and achieving $(x,y) = (1,1)$ as he would by choosing any $s < 1$ and achieving $(x,y) = (p(s),s)$. In other words, the function $p_1(s)$ gives the maximum level of cooperation that would deter Row from reducing his initial choice from 1 to $s < 1$; $1 - p_1(s)$ gives the minimum amount of retaliation necessary to deter Row.[5]

The detailed properties of Column's retaliation function $p_1(s)$ will be described next. They are analogous to those of Row's function $q_1(t)$. It is easy to verify that $p_1(s)$ is a continuous, strictly increasing, convex function that satisfies $p_1(0) = r_3$ (creates indifference when preemption is maximal) and $\lim_{s \to 1} p_1(s) = 1$ (creates indifference as preemption approaches zero).

The latter condition implies that Column can deter small defections by Row (s near 1) by the threat of small amounts of retaliation [$p_1(s)$ near 1]. This small level of threatened retaliation is, of course, a minimum; the threat of greater retaliation will also serve to deter Row. Nonetheless, since the coordinates x and y provide a natural scale for comparison of the levels of "cooperativeness" of the two players, it is reasonable to focus on the minimum threat necessary to deter Row's aggression, $1 - p_1(s)$, in comparison to the amount of initial aggression, $1 - s$.

Starting from $(x,y) = (1,1)$ at the deterrence equilibrium, we ask, for each possible level of aggression or preemption, what minimum level of retaliation will be required to deter a player from preempting. As a tool to study this question, we shall define the *retaliation-to-provocation (RP) ratio* to measure the relative deviation from proportionality, or tit-for-tat, that will just deter a potential aggressor. An RP ratio greater than 1 would indicate that one's retribution must be more punishing than an opponent's aggression, whereas a ratio less than 1 would indicate that one's retribution can be less punishing than the aggression.

Assume that both players are at DE and Row switches to strategy $s < 1$; his provocation or aggression would be the difference $\Delta = 1 - s$. Because $p_1(s)$ is Column's level of cooperativeness just sufficient to make Row's aggression profitless, $1 - p_1(s) = 1 - p_1(1 - \Delta)$ is the minimum retaliation required to deter Row. Thus, the RP ratio for Column is

$$\text{RP}(\Delta) = \frac{1 - p_1(1 - \Delta)}{\Delta} = \frac{1 - r_3}{r_3 - r_2 + \Delta(1 - r_3 + r_2)}$$

from the formula for $p_1(s)$ given in section 6.3. It is easy to verify that $\text{RP}(\Delta)$ is a continuous, strictly decreasing, convex function satisfying $\text{RP}(1) = 1 - r_3$ when the provocation is maximal and

$$\lim_{\Delta \to 0} \text{RP}(\Delta) = \frac{1 - r_3}{r_3 - r_2}$$

when the provocation approaches 0. Observe that $\text{RP}(1)$ is always less than 1, so rational deterrence of *maximal* preemption or aggression *never* requires the threat of equally strong retaliation in the Threat Game.

The minimum retaliation necessary to deter an opponent is less than the provocation whenever $\text{RP}(\Delta) < 1$. This is always true for large Δ since $\text{RP}(1) < 1$ and $\text{RP}(\Delta)$ is decreasing. Moreover, it is true for *every* Δ when the limit of $\text{RP}(\Delta)$ at $\Delta = 0$ does not exceed 1, which occurs if and only if

$$r_3 \geq \frac{1 + r_2}{2} \tag{6.1}$$

It follows from the definition of RP(Δ) that if condition (6.1) fails, RP(Δ) < 1 precisely when

$$\Delta > \Delta^* = \frac{1 + r_2 - 2r_3}{1 + r_2 - r_3} \tag{6.2}$$

What do these conditions signify? Clearly, condition (6.2) says that whatever the payoffs in the underlying game of Chicken are, a player does not have to threaten retaliation greater than the provocation *if the provocation is sufficiently high.* Figure 6.3 provides an example: If $r_4 = 1$, $r_3 = 1/2$, $r_2 = 1/4$, and $r_1 = 0$, (6.1) fails and (6.2) specifies $\Delta > 1/3$. The upper curve in figure 6.3 shows that for any provocation greater than 1/3, Column can deter Row by threatening retaliation less than Row's aggression in this example.

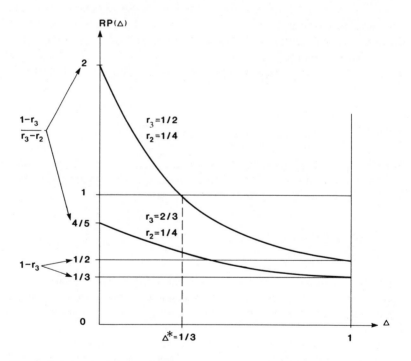

Key: Δ = level of provocation
        Normalization: $0 = r_1 < r_2 < r_3 < r_4 = 1$

Figure 6.3   Retaliation-to-provocation ratio [RP(Δ)] for two examples.

Column can *always* (that is, for any provocation $\Delta > 0$) choose less-than-proportionate retaliaion when condition (6.1) is met, which is not the case with the previous example. However, if r_3 were to change from 1/2 to 2/3 while the other payoffs remained the same, (6.1) would be satisfied [because $r_3 = 2/3 > (1 + r_2)/2 = 5/8$]. Thus, to deter Row, Column in this case need never threaten retaliation greater than the aggression (see lower curve in figure 6.3).

If condition (6.1) is not satisfied [$r_3 < (1 + r_2)/2$], then $\text{RP}(\Delta) \geq 1$ when condition (6.2) also fails. Let Δ^* be the level of provocation at which equality holds in (6.2); at $\Delta = \Delta^*$, exact tit-for-tat (i.e., retaliation equals provocation) will be just sufficient to deter an opponent. This is the *crossover point*, below which ($\Delta < \Delta^*$) retaliation must be greater than the provocation, and above which ($\Delta > \Delta^*$) it can be less, to deter aggression.

Suppose for the moment that r_2 is held fixed and r_3 is allowed to vary. Differentiation of Δ^* yields

$$\frac{d\Delta^*}{dr_3} = -\frac{1 + r_2}{(1 - r_3 + r_2)^2}$$

which is always negative. Thus, as the benefits of cooperation, r_3, increase, Δ^* decreases, indicating that the threshold below which greater-than-proportionate retaliation is called for in deterring an opponent drops as the rewards the opponent derives from compromise increase. It is therefore in the interest of one side to increase the advantages of cooperation to the other side, perhaps through certain inducements, it it wants to lower the level of its threats and, more specifically, the crossover point below which it must threaten greater-than-proportionate retaliation.

It is significant, we think, that it is against *lower* levels of provocation that a player may need to threaten proportionately *higher* levels of retribution. Of course, in some instances the minimal level to deter always stays below the provocation level (examples of both cases are shown in figure 6.3). Even when the retaliation called for starts off above the provocation level, however, it eventually dips below this level as the level of aggression rises.

The lesson in all this seems to be that minimal deterrence in the Threat Game requires threats of proportionately more punishing retaliation to deter lower levels of aggression. Put another way, to deter "salami tactics" requires the threat of a relatively strong response against an opponent who thinks about doing the "slicing."

As the level of aggression rises, more severe retaliation must be threatened, but not in proportion to the increased aggression. In the cases shown in figure 6.3, for example, threats of retaliation of only one-third and one-half of maximum aggression are sufficient to deter the aggressor. At minimal levels of aggression, by contrast, retaliation starts out at the ratio 4/5 in the first case, and at twice the provocation level in the second case.

6.5 CONCLUSIONS

We began this chapter by quoting some seemingly contradictory Biblical prescriptions related to deterrence and punishment. In Roman times, especially in the late Empire, tit-for-tat in fact became incorporated into the penal code as *lex talionis*, or the law of retaliation (see Berger, 1953, p. 730)

The notion of proportionality embedded in this law has long been considered appropriate in many societies, though in practice the punishment wreaked by some societies against their supposed enemies has been much harsher and sometimes unsparing. For example, Stalinist purges, including the Gulag, that began in the 1930s are now estimated to have killed 15–16 million Soviets (*New York Times*, August 6, 1985, p. A3).

In marked contrast to Stalin, Woodrow Wilson counseled lenient treatment of ememies, though Wilson's famous 14 points and his idealism were thwarted by allies—and later the U.S. Senate—who took a more vindictive view of Germany's aggression in World War I. More recently, McGeorge Bundy recommended a reduced response on the part of the United States if the Soviet Union should ever introduce nuclear weapons into a superpower confrontation (see Mohr, 1983). In a posthumous work, Kahn (1984) cleverly used Gedanken experiments to probe the meaning of tit-for-tat, suggesting in several scenarios involving the possible use of nuclear weapons that *lex talionis* is by no means an outmoded concept today.

It is important to bear in mind that the paramount reason for threatening to retaliate against a potential aggressor is to deter untoward acts in the first place. When these threats fail, one often faces the agonizing choice of whether or not to carry out a threat that may well be as costly to oneself as to the aggressor.

In such a situation, the usual rationale for retaliating is to bolster one's reputation for toughness and thereby deter future aggression. British Prime Minister Margaret Thatcher apparently made this calculation in responding to Argentina's invasion of the Falkland islands in 1982 by dispatching the British fleet. This conflict was very damaging to both sides, but Britain's successful invasion left little doubt about her resolve in future territorial disputes, such as might occur over Gibraltar.[6]

If Britain's response is viewed as disproportionate to the value of the Falklands yet successful for other reasons, one might ask when, if ever, retaliation should deviate from proportionality. Consider the case of Israel, which has frequently resorted to large-scale reprisals against terrorists. Its record, most recently in Lebanon, has been decidedly mixed. And where would we be if, by accident or design, a single nuclear weapon struck a major American city and the United States retaliated massively? A possible nuclear winter seems too high a price to pay for the sake of maintaining a credible deterrent, especially if that credibility can never again be tested.

Our model of deterrence based on the Threat Game suggests, at least in some cases, the following guideline: a more-than-proportionate response against relatively minor aggression, a less-than-proportionate response against relatively major aggression. The precise quantitative levels—and the crossover point at which "more" becomes "less"—depend on the payoffs of the underlying game of Chicken. However, if the payoffs to the players at the compromise outcome (r_3,c_3) are relatively high, then even at low levels of preemption a less-than-proportionate response may be sufficient for deterrence.

Historically, we will never know whether a strong policy of resistance against Hitler's early incursions would have prevented World War II, nor can we predict that after a limited nuclear first strike by one superpower a diminished response on the part of the other will prevent World War III. There are, nonetheless, good rational reasons to believe that an effective deterrent may be one in which the level of retaliation is tailored more or less—but not strictly—to the level of aggression. By threatening to hit relatively hard when the provocation is small, and backing off somewhat when large-scale conflict might prove catastrophic, one may at the same time discourage salami tactics and prevent all-out preemption.

Our model that supports this conclusion by making retaliation a function of the provocation of the more preemptive player could, in principle, be extended to a series of further moves and countermoves by the players, possibly including asymmetries of information and power.[7] In chapter 7 we use the Threat Game, commencing play from a different starting position, to analyze how a crisis, in which conflict escalates and tension rises, may be stabilized.

The prospect of stabilizing cooperation in such intractable games as Chicken and Prisoners' Dilemma is an alluring one. Yet to do so by threats of retaliation that are unnecessarily provocative is neither rational nor wise. Instead of creating an inflammatory situation by inappropriate or incredible threats that carry possibly apocalyptic consequences, especially in the case of nuclear conflict, we suggest that the quantitative guidelines of our model may provide a better strategic alternative.

It is worth stressing that these are guidelines and not detailed policy prescriptions, for we know of no practical procedures for precisely matching threats to the real-life aggression they are intended to deter. Nonetheless, we believe that the Threat Game provides a very useful framework for thinking about better and worse strategies for using threats to curb, if not prevent, preemption in certain kinds of conflicts, including nuclear.

APPENDIX

In this appendix, all undominated equilibria of the Threat Game are derived, and then all other equilibria are listed.

Since our search for equilibria is subdivided according to the final position at equilibrium, consideration of how the various final positions can arise is essential. Using the rule

$$(x,y) = \begin{cases} (p(s),s) & \text{if } s < t \\ (t,s) & \text{if } s = t \\ (t,q(t)) & \text{if } s > t \end{cases}$$

it is not difficult to check that
(a) $(x,y) = (1,1)$ occurs iff $s = t = 1$
(b) $(1,y)$, where $y < 1$, occurs iff $s = y$, $t > y$, and $p(y) = 1$
(c) (x,y), where $y < x < 1$, occurs iff either $s = y$, $t > y$, and $p(y) = x$
  or $t = x$, $s > x$, and $q(x) = y$
(d) (x,x), where $x < 1$, occurs iff either $s = x$, $t > x$, and $p(x) = x$
  or $t = x$, $s > x$, and $q(x) = x$
  or $s = t = x$

The case $(x,1)$, where $x < 1$, is analogous to (b), and the case (x,y), where $x < y < 1$, is analogous to (c).

We begin to search for equilibria by assuming that the final position is $(1,1)$ and identifying those strategy combinations that satisfy (a) and are in equilibrium. Assume that Column chooses a strategy with $t = 1$. Then Row can achieve a final position

$$(x,y) = \begin{cases} (1,1) & \text{if } s = 1 \\ (p(s),s) & \text{if } s < 1 \end{cases}$$

Now if $0 \le s < 1$,

$$P_R(1,1) = r_3 \ge P_R(p,s) = p + sr_2 - sp(1 - r_3 + r_2)$$

iff

$$p = p(s) \le \frac{r_3 - sr_2}{1 - s(1 - r_3 + r_2)} = p_1(s)$$

It is easy to verify that $p_1(s)$ satisfies $0 < p_1(s) < 1$ for $0 \leq s < 1$, and thus $p(s) \leq p_1(s)$, $0 \leq s < 1$, is a necessary condition for an equilibrium with final position $(1,1)$. Analogously,

$$q(t) \leq q_1(t) = \frac{c_3 - tc_2}{1 - t(1 - c_3 + c_2)}, \qquad 0 \leq t < 1$$

is also a necessary condition. It is not difficult to show that these necessary conditions are also sufficient, so that a strategy combination is an equilibrium with final position $(1,1)$ iff it satisfies

$$s = 1, \; q(t) \leq q_1(t); \qquad t = 1, \; p(s) \leq p_1(s) \tag{DE}$$

Payoffs at any equilibrium of the DE family are (r_3, c_3).

We now find all equilibria with final positon $(1,0)$. By (b), we have $s = 0$, $t > 0$, and $p(0) = 1$ at such an equilibrium. Given that Row chooses $s = 0$, Column can achieve the final position $(x, 0)$ for any x by choosing t and $p(0)$ appropriately. But $P_C(x,0) = xc_2$, so Column prefers $x = 1$, which he achieves iff he picks $t > 0$ and $p(0) = 1$.

Now suppose that Column picks some $t > 0$ and some $p(s)$ satisfying $p(0) = 1$. If $t < 1$, Row can attain

$$(x, y) = \begin{cases} (1,0) & \text{if } s = 0 \\ (p(s), s) & \text{if } 0 < s < t \\ (t, t) & \text{if } s = t \\ (t, q(t)) & \text{if } s > t \end{cases}$$

If $t = 1$, only the first three possibilities can arise. Observe that

$$P_R(x,y) = xyr_3 + x(1-y)(1) + (1-x)yr_2 + (1-x)(1-y)(0)$$

so that Row's payoff is a weighted average of r_3, 1, r_2, and 0 and is therefore maximized when the weight at 1, $x(1-y)$, is maximized. If either $0 < t < 1$ or $t = 1$, any choice of s other than $s = 0$ makes either $x < 1$ or $y > 0$; hence, $s = 0$ is Row's unique best response.

We have shown that a strategy combination is an equilibrium with final position $(1,0)$ iff it satisfies

$$s = 0; \qquad t > 0, \; p(0) = 1 \tag{PE_R}$$

Payoffs at any equilibrium in the PE_R family are $(1, c_2)$.

Similarly, a strategy combination is an equilibrium with final position $(0,1)$ iff it satisfies

$$s > 0, \; q(0) = 1; \qquad t = 0 \tag{PE_C}$$

Payoffs are always $(r_2,1)$.

We now locate those equilibria with final position $(1,y)$, where y satisfies $0<y<1$. By (b), an equilibrium of this type must have $s=y$, $t>y$, and $p(y)=1$. Suppose first that $s=y$ is fixed and observe that, by appropriate choice of a strategy, Column can achieve any final position (x,y) with $0\leq x\leq 1$. Thus, a necessary condition is

$$P_C(1,y)\geq P_C(x,y), \qquad 0\leq x\leq 1$$

Since

$$\frac{\partial}{\partial x}P_C(x,y)=c_2-y(1-c_3+c_2)\geq 0 \qquad \text{iff} \qquad y\leq\frac{c_2}{1-c_3+c_2}=y^*$$

a necessary condition for this equilibrium is $y\leq y^*$. (Note that $y^*<1$.)

Now fix y such that $0<y\leq y^*$, and suppose that Column's strategy satisfies $t>y$ and $p(y)=1$. Row can achieve

$$(x,y)=\begin{cases} (t,q(t)) & \text{if } s>t \\ (t,t) & \text{if } s=t \\ (p(s),s) & \text{if } s<t,\ s\neq y \\ (1,y) & \text{if } s=y \end{cases} \qquad (1)$$

and, of course, $P_R(1,y)=1-y+yr_3$. By playing $s\geq t$, Row can achieve any final outcome of the form (t,u) for any $u\ \varepsilon\ [0,1]$. Therefore a necessary condition is

$$P_R(1,y)\geq \max_u P_R(t,u)=\begin{cases} P_R(t,0) & \text{if } t>x^* \\ P_R(t,1) & \text{if } t\leq x^* \end{cases}$$

since

$$\frac{\partial}{\partial u}P_R(t,u)=(1-r_3+r_2)\ (x^*-t)$$

where $x^*=r_2/(1-r_3+r_2)$. If $t\leq x^*$, our condition is

$$1-y+yr_3\geq P_R(t,1)=tr_3+(1-t)r_2$$

which holds since $tr_3+(1-t)r_2<r_3<(1-y)(1)+yr_3$. If $t>x^*$, the condition becomes

$$1-y+yr_3\geq P_R(t,0)=t$$

In summary, the necessary condition derived from consideration of possible responses $s\geq t$ is either $t\leq x^*$ or $t\leq 1-y+yr_3$. But because

$x^* < r_3 < 1 - y + yr_3$, this condition is simply $t \leq 1 - y + yr_3$. Recall also that t was chosen so that $t > y$. Thus, what is necessary for the choice of $t < 1$ is $y < t \leq 1 - y + yr_3$; such a choice is possible iff $y < 1 - y + yr_3$, or $y < 1/(2 - r_3)$.

Consider for a moment the possibility that $t = 1$. Then only the last three possibilities on the right-hand side of (1) can arise; a necessary condition that Row prefer $s = y$ to $s \geq t$ is simply

$$1 - y + yr_3 \geq P_R(1,1) = r_3$$

which is always true. We conclude that Row prefers $s = y$ to any $s \geq t$ iff either $t = 1$ or $y < 1/(2 - r_3)$ and $y < t \leq 1 - y + yr_3$.

Now consider, for any t such that $y < t \leq 1$, when Row prefers $s = y$ among all his strategies with $s < t$. The necessary condition is $1 - y + yr_3 \geq P_R(p(s),s)$. First suppose that $s > y$. Then

$$P_R(p,s) = p[(1-s)(1) + sr_3] + (1-p)[sr_2 + (1-s)(0)]$$
$$\leq (1-s)(1) + sr_3 < (1-y)(1) + yr_3$$

so that the condition is automatically satisfied. But if $0 \leq s < y$, $P_R(p,s) \leq P_R(1,y)$ iff

$$p = p(s) \leq p_y(s) = \frac{1 - y + yr_3 - sr_2}{1 - s(1 - r_3 + r_2)}, \qquad 0 \leq s < y$$

It is easy to verify that $p_y(s)$ is a strictly increasing convex function satisfying $0 < p_y(s) < 1$ for $0 \leq s < y$.

The conjunction of the necessary conditions derived above can also be shown to be sufficient, so that a strategy combination is an equilibrium with final position $(1,y)$, where $0 < y < 1$, iff $0 < y \leq y^*$ and

$$s = y, q(t) \text{ arbitrary}; \qquad t = 1 \text{ or } y < t \leq 1 - y + yr_3$$
$$p(y) = 1 \text{ and } p(s) \leq p_y(s) \text{ for } 0 \leq s < y \qquad \text{(LE}_R\text{)}$$

At such an equilibrium, payoffs are $(1 - y + yr_3, yc_3 + (1-y)c_2)$.

Analogously, an equilibrium with final position $(x,1)$, where $0 < x < 1$, arises iff $0 < x \leq x^*$ and

$$s = 1 \text{ or } x < s \leq 1 - x + xc_3, q(x) = 1 \text{ and } q(t) \leq q_x(t) \text{ for} \qquad \text{(LE}_C\text{)}$$
$$0 \leq t < x; t = x, p(s) \text{ arbitrary}$$

where

$$q_x(t) = \frac{1 - x + xc_3 - tc_2}{1 - t(1 - c_3 + c_2)}, \qquad 0 \leq t < x$$

Payoffs are $(xr_3 + (1-x)r_2, \; 1 - x + xc_3)$.

To this point we have identified all equilibria for which the final position (x,y) satisfies either $x = 1$ or $y = 1$. All of these are undominated equilibria. We now show that all other equilibria are dominated. First we show that if an equilibrium final position satisfies $0 < x < 1$ and $0 < y < 1$, then either $x = x^*$ or $y = y^*$. To see this, suppose that the final position is attained, as in (c), its analogue, or (d) above by the choices $s = y$, $t > y$, and $p(y) = x$. Since Column could achieve the final position (u,y) for $u \; \epsilon \; [0,1]$ by choosing $p(y) = u$, we must have that $P_C(x,y) \geqq P_C(u,y)$. Since

$$\frac{\partial}{\partial u} \, P_C(u,y) = c_2 - y(1 - c_3 + c_2) = (1 - c_3 + c_2) \; (y^* - y)$$

the maximum of $P_C(u,y)$ occurs at $u = x$ iff $y = y^*$. The argument is analogous if (x,y) is reached by the choices $t = x$, $s > x$, and $q(x) = y$; if $s = t = x$, as in (d), it follows that $x = y = x^* = y^*$.

We now dispose of the case $0 \leqq x < 1$, $y = 0$. Following (c) and (d), suppose first that $s = 0$, $t > 0$, and $p(0) = x$. Column's payoff is $P_C(x,0) = xc_2$, so that Column prefers $x = 1$, and no equilibrium with $x < 1$ can arise in this way. Next suppose that $t = x$, $s > x$, and $q(x) = 0$. Again, Column receives $P_C(x,0) = xc_2$, whereas Column could have gained $P_C(1,s) = sc_3 + (1-s)c_2 \geqq c_2 > xc_2$ simply by choosing $t = 1$ and, if $s < 1$, $p(s) = 1$. Finally, if $x = 0$ and $s = t = 0$, both players receive their worst possible outcome, and either would gain by playing cooperatively. We conclude that no equilibrium can possibly result in a final position $(x,0)$ with $0 \leqq x < 1$. Similarly, the final position $(0,y)$ with $0 \leqq y < 1$ cannot be in equilibrium.

We have now shown that any equilibrium not in DE, PE_R, PE_C, LE_R, or LE_C must have either $0 \leqslant x < 1$ and $y = y^*$, or $x = x^*$ and $0 \leqslant y < 1$.

Now consider any equilibrium with final position (x,y^*), where $0 \leqslant x < 1$. Payoffs at this equilibrium are $(P_R(x,y^*), \, y^*)$. Since $y^* < c_3$, both players receive less than at DE iff $P_R(y^*,x) \leqq r_3$, which is equivalent to

$$x \leqq l_r^* = \frac{r_3 - y^* r_2}{1 - y^*(1 - r_3 + r_2)}$$

Note that $x^* < l_r^* < 1$. Thus, for $x \leqq l_r^*$, an equilibrium with final position (x,y^*) is dominated by DE.

Now observe that there is an equilibrium with final position $(1,y^*)$ in the LE_R family; it can always arise through the choices $s = y^*$, $t = 1$, and $p(y^*) = 1$, provided $p(s) \leqq p_{y^*}(s)$ for $0 \leqq s < y^*$. At this equilibrium, payoffs are $(P_R(1,y^*),y^*)$. Since

$$\frac{\partial}{\partial x} P_R(x,y^*) = 1 - y^*(1 - r_3 + r_2) > 0$$

Row prefers final position $(1,y^*)$ to any final position (x,y^*) for $0 \leqq x < 1$.

It follows that any equilibrium with final position (x,y^*), with $0 \leqq x < 1$, is either strongly dominated (if $x < l_r^*$) because both players prefer DE, or weakly dominated (if $1 > x \leqq l_r^*$) because Row prefers the line equilibrium LE_R with final position $(1,y^*)$ and Column is indifferent. Similarly, every equilibrium with final position (x^*,y) with $0 \leqq y < 1$ is dominated.

For completeness we list all the dominated equilibria with final position (x,y) satisfying $0 \leqq x < 1$ and $0 \leqq y < 1$. If $x^* > y^*$,
(i) If $1 > y \geqq l_c^*$: $s = 1$, $q(x^*) = y$, and $q(t) \leqq q_*(t)$; $t = x^*$, $p(s)$ arbitrary
(ii) If $1 > y > h_c^*$: $x^* < s \leqq P_C(x^*,y)$ $q(x^*) = y$, and $q(t) \leqq q_*(t)$; $t = x^*$, $p(x)$ arbitrary
(iii) If $1 > x \geqq l_r^*$: $s = y^*$, $q(t)$ arbitrary; $t = 1$, $p(y^*) = x$, and $p(s) \leqq p_*(s)$
(iv) If $1 > x \geqq x^*$: $s = y^*$, $q(t)$ arbitrary; $x^* \leqq t \leqq P_R(x,y^*)$, $p(y^*) = x$, and $p(s) \leqq p_*(s)$
(v) If $1 > x > k_r^*$: $s = y^*$, $q(t)$ arbitrary; $y^* < t < x^*$ and

$$t \leqq \frac{P_R(x,y^*) - r_2}{r_3 - r_2}, \qquad p(y^*) = x, \qquad \text{and} \qquad p(s) \leqq p_*(s)$$

where the following definitions (for Row, and their analogues for Column) are assumed:

$$l_r^* = \frac{r_3 - y^* r_2}{1 - y^*(1 - r_3 + r_2)}, \qquad h_r^* = \frac{y^*(1 - r_2)}{1 - y^*(1 - r_3 + r_2)},$$

$$k_r^* = \frac{r_2 - y^*(2r_2 - r_3)}{1 - y^*(1 - r_3 + r_2)}$$

$$q_*(t) = \frac{P_C(x^*,y) - tc_2}{1 - t(1 - c_3 + c_2)}, \qquad 0 \leqq t < y$$

When $y^* > x^*$, the equilibria are similar. When $x^* = y^*$, equilibria (i) and (iii) reappear, as well as equilibria like (ii) for $1>y>x^*$, like (iv) for $1>x>x^*$, and

(vi) $s=x^*$, $q(t) \leqq q_*(t)$; $t=x^*$, $p(s) \leqq p_*(s)$

NOTES

This chapter is drawn from Steven J. Brams and D. Marc Kilgour, Optimal threats, *Operations Research* 35, no. 4 (July–August 1987). Copyright © 1987 by the Operations Research Society of America. Reprinted by permission of the Operations Research Society of America.

1 The passages cited here are taken from *The Torah: A Modern Commentary* (New York: Union of American Hebrew Congregations, 1981). On retribution in the Bible more generally, see Brams (1980).

2 Axelrod (1984). Axelrod's results, along with earlier work on repeated play of Prisoners' Dilemma, were briefly summarized in chapter 2, note 5. Since Axelrod's work, some of which appeared in article form in 1980–81, a number of variations on tit-for-tat and extensions of his model have been proposed, including the following: Behr (1981); Goodin (1984); Parks (1985); Molander (1985); Bartholdi, Butler, and Trick (1986); McGinnis (1986); Witt (1986); Bendor (1987); Dacey and Pendegraft (1986); and Hirshleifer and Martinez Coll (1987).

3 Axelrod's analytic model has also been applied to Chicken, and tit-for-tat proved quite robust in this game, too; see Lipman (1986). However, tit-for-tat is decidedly not optimal in the asymmetric "Truth Game" analyzed in Brams (1985b, ch. 4); this model is based on Brams and Davis (1987). A variant of the Truth Game is analyzed in chapter 8 using a different model. As will be seen, (probabilistic) tit-for-tat may be one component of an optimal mixed strategy in these games but is not, in general, optimal by itself.

4 The four weights so obtained are non-negative and sum to 1, but they do not constitute "barycentric coordinates." This is because they do not provide a unique representation for every point of the square; uniqueness is preserved only if three points (rather than the four corner points in our model) are used to define coordinates in the 2-space of the unit square.

5 If the Threat Game were based on Prisoners' Dilemma rather than Chicken, then $p_1(s)$ for this new game,

$$\bar{p}_1(s) = \frac{r_3 - r_2 + s r_2}{1 - r_2 - s(1 - r_3 + r_2)}$$

would give the threshold threat, $1 - \bar{p}_1(s)$, that supports a Nash equilibrium analogous to the deescalation equilibrium in the Deterrence Game (chapter 2). The properties of $\bar{p}_1(s)$ are generally similar to those of $p_1(s)$. In this new game, there is exactly one other Nash equilibrium; it is analogous to the escalation equilibrium in the Deescalation Game.

6 Game-theoretic analyses of the Falkland crisis can be found in Sexton and Young (1985); Zagare (1987, pp. 48–49 and 56–62); Thomas (1987); Bennett (1987); and Wang, Hipel, and Fraser (1986).

7 Formal concepts of power (moving, staying, and threat) and information (omniscience and partial omniscience) are developed and applied in a philosophical-theological context in Brams (1983). The "theory of moves" that subsumes these concepts and their interrelationships is applied to the analysis of deterrence in different real-world conflicts in Brams and Hessel (1984); Brams (1985a, chs. 6 and 7; 1985b, ch. 2); Zagare (1987); and Kilgour and Zagare (1987). See also Kilgour, De, and Hipel (1986).

7

Crisis Stability

7.1 INTRODUCTION

Prominent in the lexicon of nuclear strategists is the notion of "crisis stability." Roughly speaking, this is a state of affairs in which neither side in a two-party conflict has a first-strike advantage that would give it the incentive to preempt the other, especially in a crisis in which tensions are high and distrust is rampant.[1]

In this chapter we interpret this concept broadly to include *any* stabilizing forces—not just the lack of a first-strike advantage—in a crisis. For example, if each side can threaten devastating retaliation after a first strike because a substantial portion of its weapons can survive, this second-strike ability may well deter a first strike and therefore be crisis-stabilizing, even though each side can still do better striking first than striking second.

Crisis stability is therefore the structural feature of conflicts that, even after a conflict has escalated to crisis proportions, enables decision makers to prevent the crisis from exploding. The principal means of avoiding an explosion is to ensure that neither side has an incentive to preempt the other in the crisis.[2]

Nuclear deterrence between the superpowers, as long as it is assured by a second-strike capability, would appear to provide the necessary insurance. What worries the political leaders of each superpower considerably more than a "bolt from the blue" by the other superpower is the possible escalation of a conventional conflict, such as might occur in the Middle East or Western Europe, into a crisis that involves serious threats to their allies or even to their own security. Insofar as the ability to stabilize such a conflict lies outside their control, crisis stabilization, at least on their part, is jeopardized.

Presumably, decision makers in an escalatory situation would prefer to be able to damp down a conflict, or at least contain it, so that a crisis cannot erupt into full-fledged war in which everybody may suffer egregiously. This is an especially frightening problem in a confrontation between nuclear powers. But even in conventional conflicts, such as that between Iran and Iraq today, the human casualties and material damage on both sides may be horrendous.

The ability to stabilize a crisis or contain a conflict, however, may be a mixed blessing. For once the protagonists recognize that disaster may be escapable, they may be more willing to take risks that they would avoid if escape from disaster were impossible, as we showed in chapter 4.

Put another way, crisis stability may encourage provocative behavior, whereas an inability to stabilize crises may induce more cautious choices. In the Cuban missile crisis of October 1962, for example, the leaders of the superpowers assiduously eschewed making explicit nuclear threats as the crisis heightened (Morgan, 1986, p. 178), perhaps in part because each side's second-strike capability at the time was not as great as it is today (notwithstanding the greater vulnerability today of each side's land-based ballistic missiles). In the absence of stabilizing forces, the situation was already extremely delicate, so neither side wanted to provoke the other into preempting with nuclear weapons.

On balance, however, crisis stability is surely desirable: While it affords each side greater leeway to be provocative, which itself may be upsetting, it makes escalation of a crisis decidedly less likely. Because stability is an eminently game-theoretic concept, it is appropriate to model crisis stability using a game. Accordingly, we begin our analysis with the Threat Game used to analyze optimal threats in chapter 6. The deterrence equilibrium (DE) in this game, as we will show, may be difficult to sustain if a crisis pushes the players into bellicose postures.

How they can prevent their positions from deteriorating further in a crisis is one theme of our analysis.[3] Specifically, after indicating what crisis stabilization entails, we will derive threat functions or lines, showing how the players in a crisis can deter further escalation, and then spell out conditions under which this task is most trying.

Unfortunately, under certain conditions there are crisis points that cannot be stabilized without an escalation in threats by the players: The threats of reprisal that sustained cooperation originally no longer suffice to stabilize the crisis and thereby deter preemption past this crisis point. In effect, the players will already have gone too far to make a return to mutual cooperation, or even stabilization, worth their while without threat escalation.

We will show that a game that requires threat escalation to stabilize near the cooperative outcome is one in which neither player can benefit from conflict in a crisis. In such a game, each player does relatively well at the cooperative outcome, lessening the temptation to defect. However, if the players should drift away from this outcome in a crisis, more dire threats are needed to prevent further deterioration of the crisis.

By contrast, if neither player is particularly advantaged at the cooperative outcome and both can do better by escalating their conflict (to some degree), the threats required to restore the status quo do not escalate, and each player is tempted to foment a (mild) crisis. If the crisis should become severe enough to hurt both players, however, then threat escalation is required to restore the status quo.

The plausibility of this game as a model of nuclear deterrence, and as a basis for determining when deterrence can be maintained in a crisis, will also be assessed. We will then return to a discussion of the substantive interpretation of this game in light of our findings about crisis stability, particularly in superpower conflict, and conditions for maintaining it. The Cuban missile crisis is used to illustrate some of the results derived from the game-theoretic model.

7.2 DETERRENCE IN A CRISIS

In section 6.2 we assumed that the preplay position of the players in the Threat Game is (1,1). To take account of the possibility that history may not start the players off from blissful cooperation, we posit the occurrence of a crisis that transports the players to some less-than-cooperative point on the game board.

More formally, a *crisis* is an event, or series of events, that changes the position of the players in the Threat Game from (1,1) to (x_0, y_0), where either $x_0 < 1$ or $y_0 < 1$ or both—that is, at least one player has acted so that $(x_0, y_0) \neq (1,1)$. We assume for now that the new preplay position is near—but not at—(1,1), in a sense to be made precise later.

In essence, the crisis has created a situation in which the players find themselves, more or less simultaneously, more hostile toward each other than previously, perhaps because one player partially preempted and the other responded with some noncooperative countermove. Indeed, a crisis suggests at least a partial failure of deterrence, which may occur for any of a variety of reasons, including domestic political considerations, misperceptions by the players, lack of communication, and the like. The question we address here is how deterrence can be restored—what kinds of threats are needed to prevent further deterioration of the crisis.

If the players wish to return to DE at (1,1) eventually, an option for each is to try to do so immediately. By virtue of the fact that DE is a Nash equilibrium, if either player credibly announces his DE strategy, his opponent can do no better than return to his own DE strategy. Thus, if Column were to announce $t = 1$, $p(s) \leq p_1(s)$, Row would maximize his payoff by returning to $s = 1$. The desire to protect himself against possible future preemption by Column would then motivate Row to threaten retaliation as well—that is, $q(t) \leq q_1(t)$.

One problem with effecting stabilization immediately at DE is that it would necessarily involve threats against the status quo (x_0, y_0). By assumption, the game is not at $(1,1)$ after the crisis has occurred, so some intimidation would be required to return to this point. In the middle of a crisis, however, a threat of retaliating against the status quo might indeed be dangerous since it could be interpreted as moving up the escalation ladder. Moreover, to create an incentive for an opponent to return to $(1,1)$, the threatener would have to change to a cooperative position himself; this may not be easy to do in a crisis.

It may be preferable for the players to try first to stabilize the status quo (x_0, y_0), postponing further ameliorative measures for restoring the nonpreemption position at $(1,1)$ until the crisis atmosphere has cleared.[4] This might be done by indicating a temporary acceptance of the current position and using presumably milder threats to deter *further* preemption.

Such a tactic, in forestalling a deepening of the crisis by arresting it at the point where it presently is, affords the players the opportunity to buy more time. Yet it poses a new question: Is it always possible, given a preplay positon near but not at full cooperation, for one or both players to stabilize (x_0, y_0), using less provocative threats than those supporting deterrence?

If the players perceive their problem as one in which the new status quo (x_0, y_0) brought on by the crisis is the best (i.e., most cooperative) position they can hope for in the short term, then we may consider the revised Threat Game to be a game played on a reduced "game board," comprising those points that lie below and to the right of (x_0, y_0) rather than $(1,1)$ (see figure 7.1). In the appendix, this short-term deterrence problem is modeled by a Reduced Threat Game—or simply Reduced Game—wherein the preplay position is assumed to be (x_0, y_0).

It is shown that, provided that (x_0, y_0) is not too far from $(1,1)$ [specifically, $x_0 > x^* = r_2/(1 - r_3 + r_2)$ and $y_0 > y^* = c_2/(1 - c_3 + c_2)$], this game has a deterrence equilibrium, analogous to (DE), defined by

$$s = y_0, \ q(t) \le q_{x_0,y_0}(t); \qquad t = x_0, \ p(x) \le p_{x_0,y_0}(s)$$

where

$$p_{x_0,y_0}(s) = \frac{P_R(x_0,y_0) - sr_2}{1 - s(1 - r_3 + r_2)}, \qquad 0 \le s < y_0$$

and

$$q_{x_0,y_0}(t) = \frac{P_C(x_0,y_0) - tc_2}{1 - t(1 - c_3 + c_2)}, \qquad 0 \le t < x_0$$

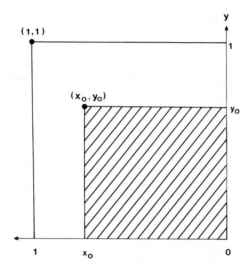

Figure 7.1   Reduced game board at (x_0,y_0).

It is this reduced deterrence equilibrium (RDE) at (x_0,y_0), to be discussed in section 7.3, that serves as the cornerstone of our model of crisis stabilization.

7.3 CRISIS STABILIZATION

We now examine more closely the properties, feasibility, and strategic implications of RDE. A few initial observations help to set the stage. First, notice that $p_{1,1}(s)=p_1(s)$; similarly, $q_{1,1}(t)=q_1(t)$. This means that as (x_0,y_0) approaches $(1,1)$, RDE at (x_0,y_0) approaches DE, the deterrence equilibrium of the Threat Game.

Other mathematical properties of RDE in the Reduced Game are important. Like $p_1(s)$ and $q_1(t)$ in the Threat Game, $p_{x_0,y_0}(s)$ and $q_{x_0,y_0}(t)$ are continuous, strictly increasing, and convex: As the levels of initial cooperation, s and t, increase, the minimal threats of retaliation, $1-p_{x_0,y_0}(s)$ and $1-q_{x_0,y_0}(t)$, necessary to deter will decrease faster and faster. In other words, players must be relatively *more* threatening at low levels of initial provocation than at high levels, sometimes threatening even greater retaliation than the provocation at these levels. The nature of such threats was analyzed in detail in chapter 6.

Perhaps the simplest way to understand all these deterrence equilibria is by means of *threat lines*. Each player threatens the other with just enough retaliation (at the threshold) to make him prefer no preemption to unilateral preemption at any level. The threat line shows how this minimum level of retaliation depends on the level of preemption.

For example, at DE in the Threat Game, Column threatens to punish any level of preemption by Row (i.e., any $s<1$) by cooperating at level no more than $p_1(s)$. Column's threat line (really a curve) at this equilibrium, called the *basic threat line*, is given by

$$x = p_1(y) \qquad \text{for } p \le y < 1$$

and is shown in figure 7.2. Similarly, Column's threat line supporting the RDE at (x_0, y_0) in the Reduced Game is given by

$$x = p_{x_0,y_0}(y) \qquad \text{for } 0 \le y < y_0$$

The upper bound, y_0, on y means that at RDE Column does not threaten to retaliate against the status quo y_0, just as Column does not retaliate against $y = 1$ at DE in the Threat Game.

In figure 7.2 we illustrate several threat lines of Column (in addition to his basic threat line) for typical RDEs. Row has a corresponding threat line at each RDE; in general, the threat line of Row supporting an RDE at (x_0, y_0) is given by

$$y = q_{x_0,y_0}(x) \qquad \text{for } 0 \le x < x_0$$

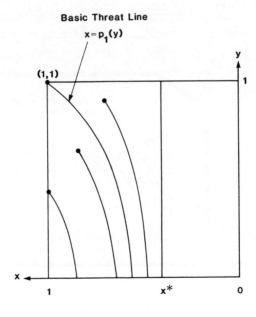

Figure 7.2   Several threat lines of Column.

Note that the threat lines supporting RDE at (x_0,y_0) always begin at (x_0,y_0). It is shown in the appendix that a player's threat lines supporting RDEs at different status quo points (x_0,y_0) never cross—they are either coincident or disjoint, as illustrated by the "parallel" threat lines of figure 7.2.

In order that it be rational for the less preemptive player to retaliate against the more preemptive player, (x_0,y_0) must be sufficiently close to $(1,1)$—that is, satisfy $x_0 > x^*$ and $y_0 > y^*$, where the lower bounds x^* and y^* are as defined in section 6.3. At higher levels of preemption, by comparison, retaliation by the less preemptive player sufficient to deter his opponent is irrational in the sense that it hurts the retaliator as well as his opponent. In effect, we consider whether a crisis can be stabilized at (x_0,y_0) in the Reduced Game when *rational* retaliatory threats by both players are possible.

We can now answer the question posed in section 7.2: When can one or both players attempt immediate crisis stabilization of a status quo point (x_0,y_0) without threat escalation. We consider two cases. In the first case, the basic threat lines of both Row and Column take the form shown in figure 7.3.

Observe that all points in the unit square near $(1,1)$ are either below Column's basic threat line or to the right of Row's (or both). Therefore, at least one player [and sometimes both, in the region of overlap between $(1,1)$ and (x',y')] can stabilize a crisis at (x_0,y_0) with the same threats meant to deter preemption at $(1,1)$.

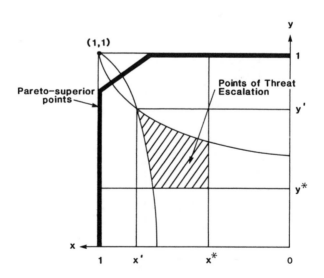

Figure 7.3 Configuration of basic threat lines in which crisis stabilization without threat escalation is always possible near (1,1).

The case illustrated in figure 7.3 occurs whenever, in the neighborhood of $(1,1)$, $x = p_1(y)$ lies to the right of $y = q_1(x)$. It is shown in the appendix that this takes place if and only if

$$\frac{1-c_3}{c_3-c_2} > \frac{r_3-r_2}{1-r_3} \tag{7.1}$$

Notice that for fixed c_2 and c_3, inequality (7.1) holds if r_3 is relatively close to r_2, and similarly for fixed r_2 and r_3. In general, (7.1) holds when r_3 and c_3 are relatively small—that is, when the players do not value the status quo very highly. For example, if $r_3 = 0.5$, $c_3 = 0.6$, and $r_2 = c_2 = 0.3$, then inequality (7.1) is satisfied.

If inequality (7.1) does not hold, figure 7.4 applies, and there are initial points close to $(1,1)$—specifically, near the main diagonal where the threat lines diverge—from which neither player can attempt crisis stabilization without escalating his threats. On the other hand, if (x_0,y_0) is sufficiently close to one of the edges of the unit square, this status quo point will be either below Column's basic threat line (near the vertical axis) or to the right of Row's (near the horizontal axis), and one player can stabilize (x_0,y_0).

A *point of threat escalation* is one that is neither below Column's basic threat line nor to the right of Row's, which in figure 7.4 means below and to the right of $(1,1)$. In figure 7.3 there are also points of threat escalation southeast of $(1,1)$, but they are not in the neighborhood of $(1,1)$. Instead, they are points below and to the right

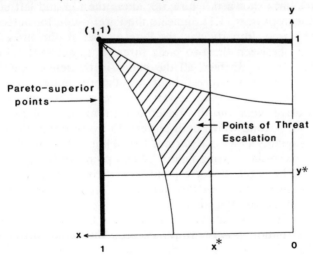

Figure 7.4 Configuration of basic threat lines in which crisis stabilization without threat escalation is not possible near (1,1).

of the *interior* point of intersection (x',y') of the two basic threat lines.

Actually, all points in the regions where the curves "flare outward" [from (x',y') in figure 7.3 and from (1,1) in figure 7.4] are points of threat escalation. It is noteworthy that, whether figure 7.3 or figure 7.4 applies, crisis stabilization is always facilitated if (x_0,y_0) falls near one of the edges of the unit square. This is because the preemptor has more to lose and is thus more vulnerable to threats. Hence, a crisis in which both players are more or less equally noncooperative is generally harder to stabilize, except when inequality (7.1) holds and the players are close to (1,1) and therefore "protected" (see figure 7.3).

In general, crisis stabilization is enhanced when the advantages of mutual cooperation in the Threat Game are substantial for the players and, in the Reduced Game, the status quo at (x_0,y_0) is not too far from (1,1). In addition, crisis stabilization is abetted if (x_0,y_0) significantly favors one player; this makes his opponent's threat of retaliation potentially more harmful and hence more likely to be taken seriously.

There is one curious aspect to crisis instability in the neighborhood of $(x,y) = (1,1)$ that suggests why players may have difficulty when the basic threat lines flare out as in figure 7.3. As shown in the appendix, the expected payoffs, $P_R(x_0,y_0)$ and $P_C(x_0,y_0)$, may actually exceed their values of r_3 and c_3, respectively, when $(x_0,y_0) \neq (1,1)$. That is, *both* players may be able to improve their payoffs at points on the game board below and to the right of (1,1), where the players are less than fully cooperative.

The dark lines shown in figure 7.3, along the top and left edges of the square except near (1,1), indicate the Pareto-superior outcomes of the Threat Game. In particular, the Pareto-superior outcomes (on the oblique line) between the two basic threat lines are better for both players than (r_3,c_3). In fact, all the points in the lens-shaped region between the two basic threat lines are Pareto improvements over (r_3,c_3), so the players would appear to have good reason to foment crises that drive them into this region and toward the oblique Pareto-superior line, beyond which both players do worse rather than better.

When inequality (7.1) is reversed, the oblique Pareto-superior line of figure 7.3 vanishes, giving figure 7.4 and points of threat escalation near (1,1). These points correspond to outcomes Pareto-inferior to (r_3,c_3); moreover, one must threaten worse reprisals than at (1,1) to make it rational to stay at them.

The level of threats required in the Reduced Game to stabilize crises at threat escalation points always exceeds the basic threats in the Threat Game that stabilize (1,1) and make it a deterrence equilibrium. Thus, threats at crisis points $(x_0,y_0) \neq (1,1)$ but close to (1,1) in figure 7.4 must be *more* severe than those at (1,1) to prevent further deterioration of the crisis.

We believe that our theoretical findings accord with reality in certain strategic situations. For instance, as the Soviets were about to install nuclear armed missiles in Cuba in October 1962, John F. Kennedy deemed his previous verbal warnings no longer adequate to alter their behavior. Claiming that his exhortations earlier had fallen on deaf ears, he escalated the conflict, imposing a blockade and contemplating even stronger actions, such as a possible air strike against the missile sites.

It seems, in such crises, that both sides may initially perceive an advantage in raising the stakes, as suggested by figure 7.3—but not to the point of an explosion. Although the tension in such crises may be unsettling, to say the least, both sides may well foresee benefits if the lid can be kept on (i.e., if they do not move past the oblique line in figure 7.3). Indeed, following the missile crisis and sobered by it, the United States and the Soviet Union quickly reached an agreement on the hot line, which they hoped would defuse future crises, and also signed the Partial Nuclear Test Ban Treaty in 1963.

These conciliatory actions surely produced a Pareto improvement in United States–Soviet relations—assuming the superpowers had moved past the Pareto-superior line and did better when they stepped back from the abyss—even if this improvement did not occur in the heat of the crisis. As in the application of many theoretical models, its "fit" may not be immediate but can be interpreted to occur over time.

Paradoxically, perhaps, a crisis that does in the end abate may have stabilizing long-term effects, even though, like the Cuban missile crisis, it may be a terrifying experience through which to live (Kavka, 1986). In terms of our model, both sides may see a short-term advantage in escalating a conflict; the increased risk of war is offset by the potential benefits of gaining a significant edge over the other side. However, as each side takes actions that bring both closer to the precipice, the dangers become more frightening and each seeks a way to step back.

Thus, while there may be an incentive to depart from the cooperative outcome in the Threat Game initially, this desire may set in motion a countervailing force. As each side becomes more threatening and conflict becomes imminent, fears arise that induce both sides to try to wind down the crisis. Then, not desiring to be so close to the precipice, each side finds it rational to retreat to (1,1)—from which, unfortunately, they will inevitably be drawn again as the memory of the last crisis fades and they become enticed once again by possibly higher payoffs.

7.4 CONCLUSIONS

It is important to emphasize that crisis stabilization is always possible in the Threat Game if either player credibly threatens retaliation at a level given by the deterrence equilibrium of this game. The new question posed in this chapter concerns simultaneous departures: If both players should, because of a crisis, find themselves at point $(x_0, y_0) \neq (1,1)$ of the unit square, can they stabilize this point using milder threats, making it a new status quo from which neither player will have an incentive to depart unilaterally?

The calculations of the Threat Game can be carried over to this Reduced Game, with (x_0, y_0) now assumed to be the preplay position. If the goal of the players is first to stabilize (x_0, y_0), and not necessarily to induce a return to $(1,1)$ immediately, then this goal cannot always be realized without escalating threats at (x_0, y_0), which may well aggravate a crisis already fraught with the danger of a conflagration.

This goal of temporary stabilization seems a sensible one for the players to aspire to if preventing a further deterioration of their positions in a crisis is a sine qua non of eventually easing out of the crisis and moving back to the *status quo ante* at $(1,1)$. Numerous crises in international politics have been abated by just such incremental steps.

For example, the eventual settlement of the Cuban missile crisis in October 1962 depended initially on the removal by the Soviets of the missiles they were about to install in Cuba. The United States provided the incentive when it escalated its verbal warnings to a blockade. Once the Soviets promised to remove the missiles, the United States agreed to lift the blockade, after which the crisis subsided.

Our game-theoretic analysis showed that crisis stabilization may not always be possible, even in the case of small departures from $(1,1)$, without threat escalation. In general, such escalation is necessary whenever both players do definitely worse in straying from $(1,1)$.

But this may not always be the case. In fact, by precipitating a crisis both players may be able to improve their payoffs, perhaps for domestic political reasons. Eventually, however, as the crisis worsens, threats may be escalated to prevent a further deterioration of relations, which may in turn push the two sides back to more conciliatory stances out of fear that their conflict could get out of hand. Yet the milder threats required to stabilize the cooperative outcome at $(x,y) = (1,1)$ may tempt the players once again to defy these threats and take risks for the higher payoffs, eventually leading to points of threat escalation where they both suffer.

The inducement to escalate threats in a crisis is heightened if the cooperative outcome is valued highly by the players, giving them less reason to depart to try to improve their lot. But even then a surprising conditon can facilitate crisis stabilization. The crisis places the players in asymmetrical positions whereby one player is substantially more antagonistic, or preemptive, than the other at (x_0, y_0). The more cooperative player is then in a position to reduce his threats, indicating (temporary) acceptance of (x_0, y_0) and thereby stabilizing it.

In the Cuban missile crisis, the Soviets initially reaped a significant advantage by their preemption, but the United States could threaten them with strong military reprisal, both of a conventional kind against Cuba and of a nuclear kind against the Soviet Union itself (Trachtenberg, 1985). More perilous than this crisis would be one in which there was rapid and simultaneous escalation on both sides and in which neither side could use an overwhelmingly greater threat against the other, perhaps because the confrontation occurred in neither's "sphere of influence."

It is perhaps fortunate that most preemptive moves made by the superpowers have been against foes that were perceived to be within their spheres of influence. The greatest danger, according to our model, exists when there is sudden and serious escalation on both sides, which seems most likely to occur when the confrontation is outside either's immediate sphere of influence, such as in the Middle East or even outer space.

It may be impossible for rational players to stabilize such a crisis without threat escalation, much less return to the *status quo ante*. We hope that a recognition that there are threat-escalation points, and that there may be no escape from them without aggravating the crisis further, will help to avert such crises.

APPENDIX

We begin by defining various "deterrence games," specifying the strategy choices available to the players, the rules for determining final outcomes, and the payoff functions. The game board of each game is its set of possible final outcomes. For example, the Threat Game (TG) has the unit square ($[0,1]^2$) as its game board, and its strategy choices, final outcome, and payoffs are as given in table 7.1 [(1_S), (1_F), and (1_P), respectively]. Note that r_2, r_3, c_2, and c_3 are constants satisfying $0 < r_2 < r_3 < 1$ and $0 < c_2 < c_3 < 1$.

Fix x_0 and y_0 satisfying $0 < x_0 < 1$ and $0 < y_0 < 1$. When the game board is reduced from the unit square to $[0, x_0] \times [0, y_0]$, the problem faced by the players can be modeled by a Reduced Threat Game (RTG). We define RTG at (x_0, y_0) to be game 2 of figure 7.5. Game 2 is played just as the original game; the only difference is in the "shrinking" of (x, y) by factors of x_0 and y_0 in (2_F) to obtain a final outcome that always lies on the RTG board.

It is possible to think of game 2, the RTG at (x_0, y_0), as being played entirely within the new game board. This is so because game 3 is equivalent to game 2, using the transformations $s' = sy_0$, $t' = tx_0$, $q'(t') = y_0 q(t'/x_0)$, and $p'(s') = x_0 p(s'/y_0)$. Although game 3 might be seen as the more natural way to reduce TG to the game board $[0, x_0] \times [0, y_0]$, we shall continue to work with game 2 for convenience. Similarly, game 4 is strategically equivalent to game 2,

Game 1: Threat Game (TG)

(1_S) Strategies: Row $\quad s \ \varepsilon \ [0, 1]$ and, if $s > 0$, $q:[0, s) \to [0, 1]$
$\qquad\qquad\qquad$ Column $t \ \varepsilon \ [0, 1]$ and, if $t > 0$, $p:[0, t) \to [0, 1]$

(1_F) Final outcome: $(x, y) = \begin{cases} (p(s), s) & \text{if } s < t \\ (t, s) & \text{if } s = t \\ (t, q(t)) & \text{if } s > t \end{cases}$

$(1p)$ Payoffs: Row $P_R(x,y) = xyr_3 + x(1-y)(1) + (1-x)yr_2 + (1-x)(1-y)(0)$
$\qquad\qquad\qquad\quad = x + yr_2 - xy(1 - r_3 + r_2)$
$\qquad\quad$ Column $P_C(x,y) = y + xc_2 - xy(1 - c_3 + c_2)$

Game 2: Reduced Threat Game (RTG) at (x_0, y_0)—Board reduction

(2_S) Strategies: As (1_S)
(2_F) Final outcome: $(x', y') = (x_0 x, y_0 y)$, where (x,y) as in (1_F)
(2_P) Payoffs: Row $P_R(x', y')$; Column $P_C(x', y')$, where P_R and P_C as in (1_P)

Game 3: RTG at (x_0, y_0)—Strategy transformation

(3_S) Strategies: Row $s' \ \varepsilon \ [0, y_0]$ and, if $s' > 0$, $q':[0, x_0 s'/y_0) \to [0, y_0]$
$\qquad\qquad\qquad$ Column $t' \ \varepsilon \ [0, x_0]$ and, if $t' > 0$, $p':[0, y_0 t'/x_0 \to [0, x_0)$

(3_F) Final outcome: $(x', y') = \begin{cases} (p'(s'), s') & \text{if } s'/y_0 < t'/x_0 \\ (t', s') & \text{if } s'/y_0 = t'/x_0 \\ (t', q'(t')) & \text{if } s'/y_0 > t'/x_0 \end{cases}$

(3_P) Payoffs: As (2_P)

Game 4: RTG at (x_0, y_0)—Payoff transformation

(4_S) Strategies: As (1_S)
(4_F) Final outcome: As (1_F)
(4_P) Payoffs: Row $\quad Q_R(x,y) = P_R(x', y')/x_0$; Column $\quad Q_C(x,y) = P_C(x', y')/y_0$,
$\qquad\qquad$ where P_R and P_C as in (1_P), and (x', y') as in (2_F)

Figure 7.5   Four deterrence games.

for the strategy sets and resulting position (x,y) are identical in both, and the payoffs in game 4 are constant multiples of the payoffs in game 2. Thus games 2, 3, and 4 are strategically identical.

We next determine the precise conditions under which games 1 and 2 are equivalent. First compare the payoff functions of games 1 and 4 for Row only; the situation for Column is analogous. From (4_P) and (2_F) it is easy to verify that

$$Q_R(1,1) = [x_0 + y_0 r_2 - x_0 y_0 (1 - r_3 + r_2)]/x_0, \qquad Q_R(0,1) = y_0 r_2 / x_0$$
$$Q_R(1,0) = 1 \qquad\qquad\qquad\qquad\qquad\qquad Q_R(0,0) = 0$$

Define $r'_2 = Q_R(0,1)$ and $r'_3 = Q_R(1,1)$. Then

$$Q_R(x,y) = xyr'_3 + x(1-y)(1) + (1-x)yr'_2 + (1-x)(1-y)(0) \qquad (5)$$

since the right side of (5) equals

$$\{xy[x_0 + y_0r_2 - x_0y_0(1-r_3+r_2)] + x(1-y)x_0 + (1-x)y[y_0r_2]\}/x_0$$
$$= [xx_0 + yy_0r_2 - xx_0yy_0(1-r_3+r_2)]/x_0 = P_R(xx_0, yy_0)/x_0 = Q_R(x,y)$$

because of (2_F) and (4_P).

Comparison of (5) with (1_P) shows that Row faces the same strategic problem in game 4 as in game 1, provided that $0 < r'_2 < r'_3 < 1$. Since

$$r'_3 - r'_2 = \frac{x_0 + y_0r_2 - x_0y_0(1-r_3+r_2)}{x_0} - \frac{y_0r_2}{x_0} = 1 - y_0(1-r_3+r_2) > 0$$

the games are strategically equivalent for Row iff $r'_3 < 1$, which is equivalent to

$$x_0 + y_0r_2 - x_0y_0(1-r_3+r_2) < x_0$$

Since $y_0 > 0$, this inequality holds iff

$$x_0 > \frac{r_2}{1-r_3+r_2} = x^*$$

Combining with the analogous conditions for Column's payoffs, we conclude that games 1 and 2 are strategically equivalent iff $x_0 > x^*$ and $y_0 > y^*$.

Now assume that $x_0 > x^*$ and $y_0 > y^*$. Our previous analysis of game 1 (TG) applies equally to RTG at (x_0, y_0) because of the strategic equivalence we have demonstrated. For game 4, the deterrence equilibrium is

$$s = 1, q(t) \leq q_1(t); \qquad t = 1, p(x) \leq p_1(s) \qquad \text{(DE)}$$

where

$$p_1(s) = \frac{r'_3 - sr'_2}{1 - s(1 - r'_3 + r'_2)}, \qquad 0 \leq s < 1$$

and $q_1(t)$ is similar. Of course, DE is identical in game 2, for (4_S) and (2_S) are the same. The transformation given above shows, after some simplification, that in game 3 DE becomes

$$s' = y_0, q'(t') \leq q_{x_0,y_0}(t'); \qquad t' = x_0, p'(s') \leq p_{x_0,y_0}(s') \qquad \text{(RDE)}$$

where

$$p_{x_0,y_0}(s') = \frac{P_R(x_0,y_0) - s'r_2}{1 - s'(1 - r_3 + r_2)}, \qquad 0 \le s' < y_0$$

and

$$q_{x_0,y_0}(t') = \frac{P_C(x_0,y_0) - t'c_2}{1 - t'(1 - c_3 + c_2)}, \qquad 0 \le t' < x_0$$

Here $P_R(x_0,y_0)$ and $P_C(x_0,y_0)$ are as given by (1_P).

Finally we show that near $(1,1)$ Column's basic threat line $(p_1(y),y)$ lies to the right of Row's basic threat line $(x, q_1(x))$ iff

$$\frac{1 - c_3}{c_3 - c_2} > \frac{r_3 - r_2}{1 - r_3} \tag{6}$$

(See figures 7.4 and 7.5 in the text.) Recall that

$$p_1(y) = \frac{r_3 - yr_2}{1 - y(1 - r_3 + r_2)}, \qquad 0 \le y > 1$$

$$q_1(x) = \frac{c_3 - xc_2}{1 - x(1 - c_3 + c_2)}, \qquad 0 \le x < 1$$

Since $\lim_{y \to 1} p_1(y) = \lim_{x \to 1} q_1(x) = 1$, both threat lines pass through $(1,1)$.

We now show that at $(1,1)$ the slope of $(x, q_1(x))$ is greater than that of $(p_1(y),y))$ iff (6) holds. Since

$$\frac{dq_1}{dx} = \frac{(c_3 - c_2)(1 - c_3)}{[1 - x(1 - c_3 + c_2)]^2}$$

we have that the limiting slope of $(x, q_1(x))$ at $x = 1$ is

$$\left.\frac{dq_1}{dx}\right|_{x=1} = \frac{(c_3 - c_2)(1 - c_3)}{(c_3 - c_2)^2} = \frac{1 - c_3}{c_3 - c_2}$$

Similarly,

$$\left.\frac{dp_1}{dy}\right|_{y=1} = \frac{1 - r_3}{r_3 - r_2}$$

so that the limiting slope of $(p_1(y),y)$ at $y = 1$ is

$$\left(\frac{dp_1}{dy}\bigg|_{y=1}\right)^{-1} = \frac{r_3 - r_2}{1 - r_3}$$

Comparison of these two slopes now yields (6).

It is not difficult to verify that any point (x,y) on Column's threat line at (x_0, y_0) satisfies $P_R(x,y) = P_R(x_0, y_0)$. (This shows that threat lines never intersect; they are either coincident or disjoint.) Now a threat line for Column can be drawn through each point of the unit square. It is easy to show that for any point (x_0, y_0) the threat line separates those outcomes that satisfy $P_R(x,y) > P_R(x_0, y_0)$ from those for which $P_R(x,y) < P_R(x_0, y_0)$. The outcomes preferred by Row are always on the side of the threat line containing $(1,0)$. These facts also apply to Column's payoff and Row's threat lines.

These observations provide a way to identify Pareto-superior outcomes. If the two threat lines from a point "flare away" (as in figure 7.4), the point is Pareto-superior to all those points below and to the right; otherwise (figure 7.3) it is not. Proceeding as in the derivation of (6) yields the sets of Pareto-superior outcomes shown in figures 7.3 and 7.4.

NOTES

This chapter is drawn from Steven J. Brams and D. Marc Kilgour, Threat escalation and crisis stability: a game-theoretic analysis, *American Political Science Review* 81, no. 3 (September 1987): 833–850.

1 We use "preemption" to mean initial attack, the implication being that each side may attack first out of fear of being attacked. This fear, as we will show, may be well grounded in the Threat Game (to be analyzed further in this chapter), especially when both sides may reap benefits simultaneously from escalation to points away from the cooperative outcome. Benefits from escalation are assumed to be related to the probability of escalation in a model that is developed and tested in Bueno de Mesquita and Lalman (1986); the benefits in this model are rooted in one player's decision-theoretic calculations, whereas in our model both players are assumed to make game-theoretic calculations, from which we derive different Nash-equilibrium strategies in crises. Game- theoretic models of crises closely related to the one discussed herein include Langlois (1987) and Stefanski (1987). A very different game-theoretic model of crisis escalation from ours, involving coalitions that recruit members and challenge an opposing coalition in some issue space, is developed in Morrow (1986). Finally, a fascinating alternative game-theoretic model of escalation, based on a dollar auction in which the two highest bidders pay, is described in O'Neill (1986) and Leininger (1987); see also Shubik (1971).

2 A more extended discussion of crisis instability can be found in O'Neill (1987), in which different efforts to formalize crisis instability are reviewed and a game-

theoretic index of this concept is defined axiomatically and applied; see also Intriligator and Brito (1976). The analysis of a "cone of mutual deterrence" by Intriligator and Brito is somewhat akin to our geometric analysis of areas of stability and instability later and has been extended in Intriligator and Brito (1985).

3 In a noncrisis context, the "robustness" of deterrent threats has been analyzed in Brams (1985b, pp. 36–43).

4 As one droll commentator, comparing the United States and the Soviet Union to two powerful automobiles on a collision course, put it: "It's infinitely easier to put both cars into reverse once they have come to a stop." [Paul Simon, quoted in "Nuclear Freeze Debate Spawns Metaphor War," *New York Times*, April 25, 1986, p. A14.]

8

Verification

8.1 INTRODUCTION

Arms control between the superpowers seems at a dead end unless the provisions of any treaty they negotiate can be verified. In a recent assessment, one expert argued that "verification has become the most important standard against which arms control agreements—both past and prospective—are measured" (Potter, 1985, p.1), and then went on to point out that

five years ago . . . verification was a contentious issue in the arms control debate, but it was not the decisive one. Today, this situation has changed, and verification stands as the litmus test by which arms control proposals are assessed. (Potter, 1985, p.4)

Another expert expressed a similar viewpoint about the primacy of verification:

The ability of the United States to verify Soviet compliance with the provisions of arms control agreements . . . has been taken by many as a fundamental requirement for establishing that a treaty is in the interests of the United States. (Richelson, 1985, p.202)

Not surprisingly, an extensive literature on verification has developed in recent years.[1]

The formal analysis of verification, based on strategic models of rational choice, has a curious history. Twenty years ago, such analysts as Maschler and Rapoport developed rather elaborate game-theoretic models to analyze the strategic implications of different inspection procedures.[2] Then, except for some scattered and mostly unpublished reports, game theory and other formal tools of rational strategic analysis were not again applied to arms control issues until recently. We review some of this work in section 8.2; here we preview the scope and main results of this chapter.

First, as a basic model of verification we propose an asymmetric variable-sum game between an inspector and an inspectee that differs from games that have previously been analyzed in the verification literature. For reasons to be given, we believe that this game better than others captures the crucial features of the strategic conflict between the superpowers over the issue of verification.

Endowing the inspector with an ability to detect the choices of the inspectee, but only imperfectly, we first analyze the payoffs of the players when the inspector induces the inspectee to respond to the inspector's announced strategy. Inducement may be thought of as a device for one player's leading the other to do something, or deterring him from doing something, in such a way as ultimately to benefit himself.

Next we systematically analyze all Nash equilibria in this game. The dependence of equilibria and inducement outcomes on the inspector's detection probabilities—both to ascertain compliance and to uncover noncompliance—is shown.

The two solution concepts of inducement and equilibrium are then compared, and the applicability of each to practical verification issues are discussed. Our goal is to provide both a theoretical framework for viewing such issues and to indicate what kinds of strategies are appropriate for achieving outcomes that help to dispel distrust and stabilize cooperation.

Trust and stability are closely linked: Adversaries can afford to trust each other if an outcome is stable—that is, if a unilateral deviation will hurt the deviator—whether the players' choices are simultaneous (Nash equilibrium) or sequential (inducement outcomes). Thus, if the inspectee is robbed of the incentive to cheat, and the inspector of the incentive to hide or manipulate the information he collects, trust will be fostered.

Moreover, this is trust based not merely on blind faith but rather on calculations of advantage and disadvantage. In the context of arms control, we derive conditions under which it is more costly to violate than not to violate the provisions of a treaty when the players choose strategies that are in some sense stable.

Thereby we are able to connect verification to stability and ultimately to trust. In fact, the connection is two-sided: Verification can give one good reason both to trust an opponent and to be trustworthy oneself.

This is not to say that present technologies for verifying adherence to arms control agreements are adequate for all weapons systems. Immense practical problems exist in detecting the deployment of certain strategic weapons, such as cruise missiles. In addition to clarifying the qualitative notions of stability and inducement, our game-theoretic analysis provides as an important by-product a demonstration that there are quantitative thresholds, as measured by the level of detector reliability, sufficient to ensure different kinds of stable and cooperative outcomes.

8.2 THE VERIFICATION GAME

A number of different games have been proposed as models of the strategic situation facing a possible violator of an arms control agreement, whom we call Inspectee, and an actor who reacts to possible violations, whom we call Inspector. Before reviewing these, we will posit our own game, which we call the (Basic) Verification Game, and then draw some comparisons between this game and others that have been proposed as models of verification.[3]

In the Verification Game we assume that Inspectee may either comply with (C) or violate (i.e., not comply with, $\bar{C}$) an arms control agreement but that Inspectee always *claims* that he chose C. Inspector may either accept (A) or challenge (i.e., not accept, $\bar{A}$) Inspectee's stated compliance. (Later we use mixed strategies to model the possibility that Inspectee may choose any *level* of compliance or noncompliance, and Inspector any *degree* of acceptance or non-acceptance, but this complication is not necessary to describe the basic game.)

In the basic version of the Verification Game, each player has two strategies that lead to four possible outcomes at the intersection of each pair of strategy choices, as shown in figure 8.1. Unlike Prisoners' Dilemma and Chicken, which underlie all previous games that we have analyzed, the Verification Game is not symmetric: The players do not rank the outcomes along the main diagonal the same, nor are the off-diagonal outcomes (lower left and upper right) mirror images of each

	Inspector (Column)	
	Accept (A)	Challenge/ Don't accept ($\bar{A}$)
Inspectee (Row) Comply (C)	$(r_3,c_4) = (r_3,1)$	$(r_1,c_3) = (0,c_3)$	s
Violate/ Don't comply ($\bar{C}$)	$(r_4,c_1) = (1,0)$	(r_2,c_2)	$1-s$
	u	$1-u$

Key: (r_i,c_j) = (payoff to Row, payoff to Column)
r_4,c_4 = best; r_3,c_3 = next best; r_2,c_2 = next worst; r_1,c_1 = worst
$s/(1-s)$ = probability Row chooses C/$\bar{C}$
$u/(1-u)$ = probability Column chooses A/$\bar{A}$
Normalization: $0=r_1<r_2<r_3<r_4=1$; $0=c_1<c_2<c_3<c_4=1$

Figure 8.1 Basic Verification Game.

other. As is evident, the strategic choices facing Inspector and Inspectee are very different, and our game model reflects this asymmetry.

The payoff rankings in the Verification Game can be characterized by the primary (1) and secondary (2) goals of each player:

Row (Inspectee)

  1 Prefers Column *accept* his claimed compliance (two best outcomes associated with A, two worst with $\bar{A}$).

  2 Prefers to *violate* agreement (whether Column chooses A or $\bar{A}$, Row prefers $\bar{C}$).

Column (Inspector)

  1 Prefers Row *comply* (two best outcomes associated with C, two worst with $\bar{C}$).

  2 Prefers policy of *tit-for-tat* (if Row chooses C, Column prefers A; if Row chooses $\bar{C}$, Column prefers $\bar{A}$).

These goals determine for each player a lexicographic order (Fishburn, 1974). The primary goal distinguishes the two best from the two worst outcomes for each player and, given this distinction, the secondary goal orders the two best, on the one hand, and the two worst on the other.

Thus, Row's primary goal implies that his two best outcomes are in the A column of figure 8.1, and his secondary goal implies that between these two outcomes he prefers the one associated with $\bar{C}$. Therefore, Row's payoff from strategy pair $\bar{C}A$ is $r_4 = 1$, and his payoff from strategy pair CA is r_3; similarly, between his two worst payoffs in the $\bar{A}$ column (as implied by his primary goal), his secondary goal implies that $\bar{C}\bar{A}$ yields r_2 and $C\bar{A}$ yields $r_1 = 0$, where $0 = r_1 < r_2 < r_3 < r_4 = 1$.

The primary goals of each player would certainly appear plausible: Row would want his *claimed* compliance accepted, and Column would want *actual* compliance. Also, Row's secondary goal, if morally dubious, is probably realistic in many situations. For if violating an agreement were not profitable, then there would be no reason for Row not to comply, making for a trivial game with a mutually best outcome at CA.

Similarly, Column's secondary goal seems eminently defensible. First, given Row chooses C, Column's preference for A is reasonable, for why should Column not cooperate by choosing A when Row cooperates by choosing C? The second part of this tit-for-tat goal seems equally reasonable—that is, that Column would be most hurt when he (unwittingly) accepts a violation.

If the reasonableness of the goals of the players is not at issue, the question of which goal is primary and which is secondary may be more controversial. Conceivably, Row may prefer violating an agreement over having its compliance accepted by Column, and Column may prefer a policy of tit-for-tat over Row's compliance. A reversal in the priority of primary and secondary goals by each player would give rise to two new games, and a reversal by both would generate a third.

Different variable-sum games have been analyzed by other theorists. For example, Rapoport (1966, pp. 158–185) applied a number of different solution concepts to a game between an inspector and an evader, in which the primary and secondary goals of the inspector duplicated ours in the Verification Game. However, the evader had a primary goal of deceiving the inspector, not simply having his claimed compliance accepted; his secondary goal was the same as our inspector's primary goal. Maschler (1966, 1967) proposed more elaborate inspection games, involving chance and an inspector who could decide whether or not to investigate a suspicious event. In his model, Maschler assumed that the inspector could announce and commit himself to a mixed strategy, against which the potential violator would maximize his expected payoff.

Brams and Davis (1987) used this idea of inducement,[4] which was originally proposed by von Stackelberg (1934) to study price leadership, and applied it to a "Truth Game" to model superpower verification. In this game, which reverses the priority of the primary and secondary goals of the inspector but not the evader in Rapoport's game, they investigated both "inducement" and "guarantee" strategies. More specifically, they analyzed the benefits that not only the inspector but also the inspectee could realize by inducing the other player to respond to him, comparing these with the payoffs that the players could guarantee for themselves whatever the opponent's choice (analogous to minimax/maximin strategies in two-person constant-sum games). More recently, Brams and Kilgour (1986c) studied the Nash equilibrium strategies of the players in Rapoport's inspector-evader game and the Truth Game.

Fichtner (1986) compared various game-theoretic approaches to the inspection and verification of arms control agreements; in addition, he applied similar solution concepts to auditing, consumer protection, and nuclear safeguards. Avenhaus's treatment of the latter issue, in particular, makes significant use of game theory (Avenhaus, 1986); among other things, he finds optimal strategies for an inspector, with goals the same as ours in the Verification Game, to induce compliance by an inspectee, whose primary but not secondary goal is the same as ours.

There is probably no "best" game to model all aspects of verification. We have already justified the Verification Game in terms of the primary and secondary goals of the players, but it is worth exploring why its ranking is especially plausible in the case of the superpowers. Consider each of the four outcomes in turn, and assume that one superpower (Row) contemplates cheating on the arms control treaty, and the other (Column), who monitors compliance, can challenge possible violations:

$\bar{C}A$ (1,0):   A successful violation of a treaty, giving Row a substantial edge in the arms race, would certainly seem the best outcome for Row and the worst for Column.

CA   (r_3,1):   This is the best outcome for Column, for it validates the treaty, but it is definitely inferior for Row inasmuch as an unchallenged violation could give Row the edge mentioned above.

$C\bar{A}$ (0,c_3):   Column gets compliance but, without at first recognizing it, creates some embarrassment for himself by erroneously challenging Row; for Row, on the other hand, a false charge of cheating is his worst outcome, undermining the benefits of the treaty for no gain.

$\bar{C}\bar{A}$ (r_2,c_2):   A rather unsatisfactory outcome for both sides, because Column's justified challenge of violations underscores the treaty's fragile status, perhaps leading to its abrogation.

One might contend that $\bar{C}\bar{A}$ would be higher than next-worst in Column's preferences, for challenging noncompliance is certainly in Column's interest. We believe, however, that it is more realistic to suppose that Column would prefer compliance at $C\bar{A}$ to noncompliance at $\bar{C}\bar{A}$, even though compliance at the former outcome entails for Column the embarrassment of a false charge.

The reason for this preference is that the distress caused by an unjustified accusation can more easily be undone than that of an actual violation, even one that is detected and challenged. After more than forty years of conflict, both superpowers almost certainly would prefer to have their adversary adhere to SALT and other arms control treaties, even at the cost of occasionally making unsubstantiated charges, than to catch their adversary in a lie and challenge real violations.

Whether c_2 and c_3 are interchanged in the Verification Game, however, makes no difference for the rational strategy choice of Row, which is to choose $\bar{C}$. For this strategy is dominant: Whether Column chooses A or $\bar{A}$, $\bar{C}$ is strictly better than C for Row. In a game of complete information, Column would know that Row has an unconditionally best strategy choice; presuming Row to choose it, Column can do no better than choose $\bar{A}$, leading to (r_2,c_2).

This is the unique Nash equilibrium in the Verification Game, but, unfortunately for both players, it is Pareto-inferior since it is worse for both players than (r_3,1). Yet the latter outcome is not in equilibrium, because Row has an incentive to depart from it to (1,0), his best outcome.

The Pareto-inferior rational solution to this game can be circumvented if, assuming Row chooses his strategy first, Column has perfect information about Row's choice and Row knows this.[5] In this case, it is easy to show that Column would have a dominant strategy of tit-for-tat—choose A if Row chooses C, and $\bar{A}$ if Row chooses $\bar{C}$—and Row, anticipating this dominant-strategy choice, would choose C, resulting in $(r_3,1)$, a Nash equilibrium in the resulting 2×4 game.

But, of course, the superpowers in general will have only imperfect information about each other's choices when each plays the role of Inspector. How, then, can they use their imperfect detection equipment to choose optimally? In the next section, we propose two different notions of "optimal" and analyze how rational strategies in each case depend on the quality of the detection equipment in a more realistic version of the Verification Game.

8.3 INDUCEMENT IN THE VERIFICATION GAME WITH DETECTION

We assume that Column is equipped with a detector and has the option of consulting it before choosing A or $\bar{A}$. Column's detector is characterized by parameters x and y, which are *conditional probabilities*—dependent on the occurrence of prior events that follow the vertical bars in the braces below—that describe its reliability and are assumed to be known by both players:

$x = \Pr\{\text{detector signals violation} \mid \text{Row chose } \bar{C}\}$
$y = \Pr\{\text{detector signals no violation} \mid \text{Row chose } C\}$

Note that $0 \le x,y \le 1$. The detector is perfect when $x = y = 1$ and would seem worthless when x and/or y are near or at 0.[6]

We assume that Row chooses his strategy before Column chooses his. Row has two pure strategies, C and $\bar{C}$, so his mixed strategy can be represented by a single probability s:

$\Pr\{\text{Row chooses } C\} = s$

Because Column has a detector, his choices are more complicated. In addition to his two pure strategies, A and $\bar{A}$, which we treat as unconditional choices (i.e., made without consulting his detector), he has a third pure strategy of consulting his detector (D). If Column chooses D, we assume he will follow a policy of tit-for-tat by picking $\bar{A}$ if the detector signals a violation and A otherwise. Altogether, we represent Column's mixed strategy by two probabilities, t and u, where

$\Pr\{\text{Column chooses } D\} = t$
$\Pr\{\text{Column chooses } A\} = u$

Of course, Pr{Column chooses $\bar{A}$} $= 1 - t - u$, just as Pr{Row chooses $\bar{C}$} $= 1 - s$. Because the mixed strategies are probabilities,

$$0 \leq s \leq 1; \quad 0 \leq u \leq 1; \quad 0 \leq t \leq 1; \quad u + t \leq 1$$

The latter sum is the probability that Column chooses either A or D—that is, the probability that he does not choose $\bar{A}$.[7]

We have indicated the probabilities of Row's choices C and $\bar{C}$ (s and $1 - s$), and the probabilities of Column's choices A and $\bar{A}$ (u and $1 - u$), in parentheses to the right and below these strategies in figure 8.1, which represents the Basic Verificatiion Game. When Column has a detector and uses it with a probability t, figure 8.1 must be modified by changing the probability associated with Column's choice of $\bar{A}$ from $1 - u$ to $1 - t - u$. Then, combining the indicated probabilities with the probability, t, of Column's consulting his detector (D) and then choosing either A or $\bar{A}$, we obtain probabilities that each of the four possible outcomes of the game will occur:

CA: $s(u + ty)$
C$\bar{A}$: $s(1 - u - ty)$
$\bar{C}$A: $(1 - s)(t + u - tx)$
$\bar{C}\bar{A}$: $(1 - s)(1 - t - u + tx)$

For example, CA is chosen when Row chooses C (with probability s) and Column chooses either A unconditionally (with probability u) or consults his detector (with probability t) and, detecting no violation (with probability y), chooses A.

These outcome probabilities may now be combined with the payoffs that the players obtain at the outcomes (see figure 8.1) to give the players' expected payoffs in the Verification Game. To simplify notation, we distinguish the E's as before by subscripts R (for Row) and C (for Column):

$$E_R(s;t,u) = s(u + ty)r_3 + (1 - s)(t + u - tx) + (1 - s)(1 - t - u + tx)r_2$$
$$E_C(t,u;s) = s(u + ty) + s(1 - u - ty)c_3 + (1 - s)(1 - t - u + tx)c_2$$

In the appendix, we show that the maximin strategies for Row and Column—that is, the strategies that maximize their minimum expected payoffs, whatever the opponent does—are $s = 0$ and $u = t = 0$, respectively. These strategies, which are for Row always to choose $\bar{C}$ and for Column always to choose $\bar{A}$ unconditionally (and never consult his detector), give the players maximin values of r_2 and c_2, respectively. Recall that these are the payoffs to the players from choosing their Nash equilibrium strategies in the figure 8.1 game (without the possibility of detection by Column).

Later we investigate the Nash equilibria that arise when Column has an imperfect detector that he can use to try to discern Row's (prior) strategy choice. These equilibria may be different from the unique Nash equilibrium in the figure 8.1 game without detection.

First, however, we investigate how Column, by announcing and committing himself to a mixed strtegy (u,t), which we call an *inducement strategy*, can induce Row to respond to this strategy in such a way that Column maximally benefits. As we will show, Column's optimal inducement strategy depends on his detector's having detection probabilities x and y above a certain minimum. We assume, for this exposition, that both $x<1$ and $y<1$; this assumption is dropped in the appendix.

Because Row has only two pure strategies, Column can induce either one or the other by arranging that it be Row's unique best response. Column cannot induce a mixture, since no mixed strategy could consitute Row's best response unless all do, which is to say that no strategy, pure or mixed, is better than any other.

We show in the appendix that $s=0$ (always choose $\bar{C}$) is a best response for Row to any strategy of Column if

$$x(1-r_2)+yr_3 \leq 1$$

But then Column cannot receive more than c_2, whereas if $s=1$ Column cannot receive less than $c_3>c_2$ (see figure 8.1).

Obviously Column would prefer to induce $s=1$, and he can do so provided that

$$x(1-r_2)+yr_3 > 1 \tag{8.1}$$

We show in the appendix that, under this condition, Column induces $s=1$, and benefits maximally, when he chooses

$$u = \left[\frac{x(1-r_2)-1+yr_3}{x(1-r_2)-r_3+yr_3} \right]^{-} \tag{8.2a}$$

$$t = 1-u = \left[\frac{1-r_3}{x(1-r_2)-r_3+yr_3} \right]^{+} \tag{8.2b}$$

where the minus and the plus superscripts indicate values slightly less and greater, respectively, than those given in the brackets. This strategy makes $s=1$ Row's unique best response, and among all strategies that do so this one maximizes Column's expected payoff.

What are the benefits to Column of this optimal inducement strategy? It turns out that this strategy yields Column

$$E_C^* = E_C(u,t;1) = \left[\frac{x(1-r_2)-(1-y)(1-c_3+c_3r_3)}{x(1-r_2)-(1-y)r_3} \right]^-$$

and $c_3 < E_C^* < 1$. Thus, Column receives strictly more than his next-best payoff of c_3; he receives somewhat less than his best payoff of 1 because of detector unreliability.

For Row the benefits are not so great, in terms of the comparative rankings of the players, but neither are they terrible:

$$E_R^* = E_R(1;u,t) = \left[r_3\left(\frac{x(1-r_2)-1+y}{x(1-r_2)-r_3+yr_3} \right) \right]^-$$

which satisfies $r_2 < E_R^* < r_3$. In other words, Row receives strictly more than his maximim value of r_2 but strictly less than his payoff at the "cooperative" outcome $(r_3,1)$, when $s = u = 1$ and $t = 0$.

Indeed, $(r_3,1)$ gives a better payoff to *both* players than what Column can induce, so inducement is Pareto-inferior to the pure-strategy outcome $(r_3,1)$. Moreover, even these inferior inducement payoffs are unattainable unless the conditional detection probabilities x and y are sufficiently high that inequality (8.1) is satisfied. Given that (8.1) is satisfied, Column can induce either $s = 0$ or $s = 1$; by choosing u and t according to (8.2), he will induce (E_R^*, E_C^*), which is certainly better for both players than (r_2,c_2) (obtained by optimally inducing $s = 0$), even if (E_R^*, E_C^*) is Pareto-inferior to $(r_3,1)$.

It might be thought that Row could turn the tables on Column and induce Column to respond to his own mixed strategy. However, Row would have no reason to announce that he might ever choose $\overline{C}$ (violate the treaty), because this could only steer Column toward the choice of $\overline{A}$, which leads to Row's two worst outcomes. If Row is going to violate, secrecy is all-important to him.

In section 8.4 we assume that Column is not able to seize the initiative and announce his optimal inducement strategy to evoke a best response from Row. Instead we ask whether Column, again with only an imperfect detector, and Row can simultaneously (or at least in ignorance of each other's choices) choose strategies such that neither would have an incentive to depart unilaterally from his choice.

8.4 NASH EQUILIBRIUM STRATEGIES IN THE VERIFICATION GAME WITH DETECTION

Assume that Row does not respond to Column's prior choice—as under optimal inducement by Column—perhaps because Column is unable to make a credible commitment to a particular mixed strategy. Rather, suppose that both players, knowing that Column can imperfectly detect Row's prior choice, act in light of this knowledge (i.e., of the conditional probabilities x and y) and the payoffs of the Verification Game shown in figure 8.1.

Under the assumption that $x>0$ and $y>0$ (i.e., the detector has nonzero probabilities of being correct), we derive in the appendix all Nash equilibria in the Verification Game with detection. These equilibria can be classified into five distinct groupings:

I   *Cooperative equilibrium*, with payoffs of $(r_3,1)$. Since this equilibrium requires $y=1$ (perfect detection of compliance by Column), it is unrealistic, and we will not consider it further.

II   *Noncooperative equilibria*, with payoffs of (r_2,c_2). These equilibria can always (i.e., for any values of x and y) be achieved by Column's choosing $\bar{A}$ for certain and Row's choosing $\bar{C}$ for certain. If $x=1$, Column may instead consult his detector some of the time without changing the equilibrium.

In any event, these equilibria give the players only their maximin values and are dominated by all other equilibria whenever other equilibria exist (more on this point below).

III   *Constant-detection equilibrium*, with payoffs

$$(yr_3, \ xc_2 + s[c_3 + y(1 - c_2) - xc_2])$$

where $x(1 - r_2) + yr_3 = 1$. Column always consults his detector at this equilibrium and so follows a policy of tit-for-tat. This equilibrium is not very important because it occurs very rarely— only when the values of x and y precisely satisfy a linear equation.

IV   *Never-accept (unconditionally) equilibrium*, with payoffs

$$\left(\frac{yr_2r_3}{yr_3 - (1-x)(1-r_2)}, \ \frac{yc_2(1-c_3)+(1-x)c_2c_3}{y(1-c_3)+(1-x)c_2} \right)$$

where $x(1 - r_2) + yr_3 > 1$. Column either consults his detector or chooses $\bar{A}$ but never chooses A; because this equilibrium is dominated by equilibrium V (i.e., V is better for both players than IV), it will not be considered further.

V   *Never-challenge (unconditionally) equilibrium*, with payoffs

$$\left(\frac{x(1-r_2)r_3 - (1-y)r_3}{x(1-r_2) - (1-y)r_3}, \ \frac{xc_2}{(1-y)(1-c_3)+xc_2} \right)$$

where $x(1-r_2)+yr_3>1$. Column either consults his detector or chooses A but never chooses $\bar{\text{A}}$.

To summarize, we have dismissed I as unrealistic because perfect detection is unattainable, III as unimportant because it almost never occurs, and IV because it is dominated by V. We also show in the appendix that II is dominated by V (when V occurs); only the $\bar{\text{C}}\bar{\text{A}}$ form of II will be considered here, for the $x=1$ variant requires perfect detection.

This leaves II and V. If

$$x(1-r_2)+yr_3<1 \tag{8.3}$$

II is the only equilibrium. The unshaded region in the unit square below and to the left of the line shown in figure 8.2 satisfies inequality (8.3).

Thus, if the detection probabilities x and y are so low as to satisfy (8.3), only strategies associated with the noncooperative equilibrium are stable in the Verification Game. Formally, these strategies are

$$s=0; \ u=0, \ t=0$$

where $x<1$.

The situation is a little better for the players if equality holds in (8.3), for equilibrium III, which dominates II, then exists. But this occurs only *on* the line $x(1-r_2)+yr_3=1$, shown in figure 8.2.

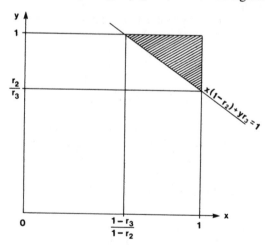

Figure 8.2   Region (shaded) in which equilibrium V dominates all other Nash equilibria and inducement is possible.

The picture is still brighter if inequality (8.3) is reversed, which is in fact inequality (8.1) in section 8.3. If x and y are sufficiently high that $x(1-r_2)+yr_3>1$, then Row and Column can obtain the payoffs given by equilibrium V. Of course, equilibria II and IV are also available in this region, which lies above and to the right of the line $x(1-r_2)+yr_3=1$ shown in figure 8.2. Presumably, however, the players will choose the dominant equilibrium in this region—that which is better for both players.

Consider the region in which equilibrium V exists and is dominant. This region, which is defined by inequality (8.1), is exactly the region in which Column can induce Row to choose $s=1$ (i.e., always comply), as shown in section 8.3.

A comparison of Column's *payoff* at equilibrium V with what he can obtain through optimal inducement is instructive. As shown in the appendix, Column does better under optimal inducement, obtaining E_C^*, than he does at equilibrium V; Row does marginally worse.

In fact, Column's optimal inducement strategy given by (8.2) is almost identical to his strategy at equilibrium V (see appendix for details). In either case, Column sometimes consults his detector, sometimes chooses A, but never chooses $\bar{A}$ without first consulting his detector. Row's equilibrium V strategy is mixed (see appendix), but of course his best response to Column's optimal inducement is pure (strategy C).

As the detection probabilities x and y approach 1—allowing for perfect detection—the payoffs to the players approach $(r_3,1)$ both under optimal inducement and at equilibrium V. In other words, the players can come closer and closer to the payoffs of the cooperative equilibrium (I) as detection improves, which is a hopeful sign in this game. Moreover, Column need never resort to choosing A unconditionally—before consulting his detector—as long as inequality (8.1) is satisfied.

Only below the line shown in figure 8.2, where inequality (8.3) is satisfied and detection is relatively poor, do the players have no equilibrium strategies other than always to choose $\bar{C}$ and $\bar{A}$. Above this line, it is advantageous for Column sometimes to consult his detector and sometimes to accept unconditionally. This mixed strategy evokes greater compliance from Row, which renders the outcome at equilibrium V stable and yields higher payoffs to both players than are available at any other Nash equilibrium.

However, whenever equilibrium V is available, it is always in Column's interest to attempt optimal inducement, for this approach would substantially improve Column's expected payoff without any significant effect on Row's. We illustrate these ideas with a numerical example: If $r_2=0.2$, $r_3=0.7$, $c_2=0.4$, $c_3=0.7$, and $x=y=0.8$, then (8.1) holds, and the players have equilibria II, IV, and V available, paying (0.2,0.4), (0.28,0.48), and (0.62,0.84), respectively. At equilibrium V, Row complies 84 percent of the time; Column consults his detector 60 percent of the time and accepts unconditionally otherwise. For optimal inducement, Column credibly commits himself to consulting his detector slightly more than 60 percent of the time and accepting unconditionally otherwise. Row's best response is to increase compliance to 100 percent, and the payoffs become $(0.62^-, 0.96^-)$; hence, Column's gain of almost 0.12 is achieved at essentially no cost to Row.

8.5 CONCLUSIONS

There are both auspicious and inauspicious implications of our analysis of the Verification Game with detection. Inauspiciously, the detection probabilities must be above threshold values before the detector can be helpful at all or inducement by Inspector is worth trying. Indeed, below these values the detector is worse than useless and should never be consulted. Instead, the players' strategies of always violating and always challenging are the only strategies in equilibrium.

Auspiciously, though the players' noncooperative strategies remain in equilibrium above the threshold values, this equilibrium is dominated by more cooperative equilibria that yield higher payoffs for both players. The most attractive of these equilibria involves mixed strategies, whereby Inspector sometimes relies on his detector, sometimes accepts (unconditionally) Inspectee's claimed compliance, but never challenges unconditionally. Inspectee's equilibrium strategy is also mixed, but between compliance and noncompliance, with the probability of compliance rising as Inspector's detector improves.

By announcing a mixed strategy to which Inspectee responds, Inspector can induce a pure-strategy response by Inspectee that raises Inspector's payoff above its greatest Nash equilibrium value (in the shaded region in figure 8.2). Inspectee is slightly hurt in this region when he is induced to comply.

Whatever the differences between the inducement and equilibrium payoffs, the key to "solving" the Verification Game lies not so much in the solution concept selected but in being in the favorable region (shaded in figure 8.2), which triggers both the possibility of inducement and Nash equilibria that are Pareto-superior to the noncooperative equilibrium. Inducement probably makes most sense if there is a third party or neutral inspector who can credibly announce acceptance conditions to which the parties to a conflict will respond. On the other hand, in a two-person game between adversaries such as the superpowers—wherein each adversary plays the role of both Inspector and Inspectee—the Nash equilibrium strategies may be more sensible because of their symmetry with respect to the players.

It is difficult to say whether the verification equipment of the superpowers is sufficiently good to place them in the favorable region. Nevertheless, the nuclear arms control agreements reached so far, though limited in scope, argue against unbridled pessimism.

Verification of the most significant agreements that the United States and the Soviet Union have achieved, beginning with the 1963 Partial Nuclear Test Ban Treaty prohibiting nuclear weapons tests in the atmosphere, in outer space, and under water, has been based on "national technical means" (NTM). The use of NTM means that each superpower relies primarily on its own surveillance of the other side's activities, using reconnaissance satellites and the like. The importance of NTM since the early 1960s should not be underestimated; in fact, without NTM, there would be no Verification Game.[8]

Perhaps we cannot be so sanguine about new and future weapons systems, including those, like cruise missiles, that are highly mobile or those, like antisatellite weapons, that are hard to detect for technical reasons. Yet, because it is in both sides' interests—whether in the role of Inspector or Inspectee—to be in the favorable region, our analysis suggests that each side's apparent obsession to conceal as much as possible is probably ill-founded.

It is worth stressing that stability, except of the noncooperative kind, is impossible unless detection thresholds are reached and surpassed. In the absence of such stability achieved by equilibrium or inducement, verification will surely fail, for the rational strategies of players—to violate and challenge—will undermine it. Thus, an overarching goal of verification, we believe, must be to develop procedures for exceeding the thresholds at which cooperative behavior is stabilized. The players have a mutual interest both in being more compliant and in developing and allowing the use of better detection methods.

APPENDIX

Maximin Strategies and Values

We begin by calculating the maximin strategies and values for the players in the Verification Game with detection, defined in section 8.3, where we showed that the expected payoffs of Row and Column are

$$E_R(s;t,u)=s(u+ty)r_3+(1-s)(t+u-tx)+(1-s)(1-t-u+tx)r_2 \qquad (1)$$
$$E_C(t,u;s)=s(u+ty)+s(1-u-ty)c_3+(1-s)(1-t-u=tx)c_2 \qquad (2)$$

Observe that E_R can be rewritten as

$$E_R(s;u,t) = (1-s)r_2 + u[sr_3 + (1-s)(1-r_2)]$$
$$+ t[syr_3 + (1-x)(1-r_2)] \tag{3}$$

Since the coefficients of u and t in (3) are non-negative, it follows that

$$\min_{u,t} E_R(s;u,t) = E_R(s;0,0) = (1-s)r_2$$

which is maximized by $s=0$. Thus, Row's maximin strategy is $s=0$, and his maximin value is r_2. These maximin results are identical to those for the Verification Game without detection.

To find Column's maximin strategy, note from (2) that

$$\frac{\partial E_C}{\partial s} = (c_3 - c_2) + u[1 - c_3 + c_2] + t[y(1-c_3) + (1-x)c_2] > 0 \tag{4}$$

From (4) it follows that

$$\min_s E_C(u,t;s) = E_C(u,t;0) = (1 - u - t + tx)c_2$$

so Column's maximin value is c_2 and his maximin strategy is $u=0$, $t=0$ (or, if $x=1$, $u=0$, t arbitrary). Again, these maximin results are the same as for the no-detection case.

Inducement by Column

Inducement occurs when Column announces in advance his (mixed) strategy, inviting Row to make his best response. In some detection games,[9] Column can do better by means of this stratagem than by the players' simultaneously selecting maximin strategies.

Since Row has only two pure strategies, Column can induce only one or the other (C or $\bar{\text{C}}$). Suppose that Column has chosen some u and t, and consider what Row's best response is. From (1) it follows that

$$\frac{\partial E}{\partial s} = -r_2 + u[r_2 + r_3 - 1] + t[yr_3 - (1-x)(1-r_2)] \tag{5}$$

Assume first that $x(1-r_2) + yr_3 \leqq 1$. Then

$$yr_3 - (1-x)(1-r_2) = x(1-r_2) + yr_3 - 1 + r_2 \leqq r_2$$

so by (5),

$$\frac{\partial E_R}{\partial s} \leqq -r_2 + ur_2 + tr_2 = -r_2 + r_2(u+t) \leqq 0$$

since $u + t \leq 1$.

We have shown that $s = 0$ is Row's best response to any strategy of Column if $x(1 - r_2) + yr_3 \leq 1$. Thus, Column cannot induce $s = 1$ in this case. This puts Column in an unfavorable position, for it is easy to verify from (2) that Column cannot receive more than c_2 if $s = 0$, and cannot receive less than $c_3 > c_2$ if $s = 1$.

Now assume that (see figure 8.2)

$$x(1 - r_2) + yr_3 > 1 \tag{6}$$

Then $yr_3 - (1 - x)(1 - r_2) > r_2$, and it is easy to show from (5) that $\partial E_R / \partial s > 0$, so Column induces $s = 1$ iff

$$t > \frac{r_2 + u(1 - r_3 - r_2)}{yr_3 - (1 - x)(1 - r_2)} \tag{7}$$

But $t \leq 1 - u$ is also required, so u must satisfy

$$\frac{r_2 + u(1 - r_3 - r_2)}{yr_3 - (1 - x)(1 - r_2)} < 1 - u$$

which can be proved to be equivalent to

$$u < \frac{x(1 - r_2) - 1 + yr_3}{x(1 - r_2) - r_3 + yr_3} \tag{8}$$

[From (6), the right side of (8) is well-defined; the numerator is positive and less than the denominator.] Hence, we have shown that if (6) holds, Column can induce $s = 1$ by choosing any $u \geq 0$ satisfying (8) and then any $t \leq 1 - u$ satisfying (7).

We now ask which, among the choices of u and t that induce $s = 1$, does Column most prefer? This choice will make Column's inducement optimal. From (3) it follows that

$$E_C(u, t; 1) = c_3 + (1 - c_3)(u + ty)$$

so Column maximizes his expected payoff by choosing $(u + ty)$ as large as possible. If $y < 1$, Column must choose

$$u = \left[\frac{x(1 - r_2) - 1 + yr_3}{x(1 - r_2) - r_3 + yr_3} \right]^-$$

$$t = 1 - u = \left[\frac{1 - r_3}{x(1 - r_2) - r_3 + yr_3} \right]^+ \tag{9}$$

If $y = 1$, any u satisfying (8) and $t = 1 - u$ is sufficient.

We have shown that if (6) fails, Column cannot induce any strategy of Row other than $s = 0$; Column then obtains at most c_2. If (5) holds, Column can induce either $s = 0$ or $s = 1$.

We next show that optimal inducement of $s = 1$ is indeed beneficial to Column. Substitution of (9) and $s = 1$ into (2) yields

$$E_C^* = E_C(u,t;1) = \left[\frac{x(1-r_2) - (1-y)(1-c_3+c_3r_3)}{x(1-r_2) - (1-y)r_3} \right]^- \quad (10)$$

[The right side of (10) is well-defined, since the donominator exceeds the numerator, and the numerator exceeds $x(1-r_2) - (1-y) > x(1-r_2) + yr_3 - 1 > 0$, by (6).]

Next we show that optimal inducement yields Column strictly more than c_3 by evaluating

$$x(1-r_2) - (1-y)(1-c_3+c_3r_3) - x(1-r_2)c_3 + (1-y)r_3c_3$$
$$= x(1-r_2)(1-c_3) - (1-y)(1-c_3)$$
$$= (1-c_3)[x(1-r_2) - (1-y)] > 0$$

It is now clear that $c_3 < E_C^* < 1$.

Finally, we determine the effects of Column's optimal inducement on Row by substituting (9), and $s = 1$, into (1):

$$E_R^* = E_R(1;u,t) = r_3 \left[\frac{x(1-r_2) - 1 + y}{x(1-r_2) - r_3 + yr_3} \right]^- \quad (11)$$

[Again it is easy to use (6) to show that the fraction on the right-hand side of (11) is well-defined.] To show that Column's optimal inducement yields Row more than r_2, we calculate

$$x(1-r_2)r_3 - (1-y)r_3 - x(1-r_2)r_2 + (1-y)r_2r_3$$
$$= (1-r_2)[x(r_3-r_2) - (1-y)r_3] > 0$$

because, by (6),

$$yr_3 - (1-x) - xr_2 > 0$$

so that

$$yr_3 - r_3(1-x) - xr_2 > 0$$

It now follows that $r_2 < E_R^* < r_3$.

Nash Equilibria

We now determine all Nash equilibria of the Verification Game with detection. To avoid trivialities, assume that the detector satisfies $x > 0$ and $y > 0$.

We first identify all those equilibria with $s = 1$. From (2),

$$E_C(u,t;1) = c_3 + (u + ty)(1 - c_3)$$

so that at any equilibrium with $s = 1$, Column must maximize $(u + ty)$ because $1 - c_3 > 0$. Thus, for an equilibrium with $s = 1$, either $u = 1$, $t = 0$, or, if $y = 1$, u is arbitrary and $t = 1 - u$.

To begin with, suppose $y < 1$. If $u = 1$, $t = 0$, (1) shows that

$$E_R(s;1,0) = 1 - s(1 - r_3)$$

But since $1 - r_3 > 0$, the choice of $s = 1$ does not maximize Row's expected payoff; therefore, there are no equilibria with $s = 1$ if $y < 1$.

We next search for equilibria with $s = 1$ under the assumption that $y = 1$. If $u = 1 - t$, then (1) shows that

$$E_R(s;1-t,t) = 1 - tx(1 - r_2) + s[tx(1 - r_2) - (1 - r_2)]$$

It follows that $s = 1$ maximizes Row's expected payoff iff

$$t \geq \frac{1 - r_3}{x(1 - r_2)}$$

which, because $t \leq 1$ is required, can be arranged iff $x \geq (1 - r_3)/(1 - r_2)$. It can be verified that these necessary conditions are also sufficient. We can now characterize all Nash equilibria with $s = 1$.

I. *Cooperative equilibrium.* There is an equilibrium with $s = 1$ iff $x \geq (1 - r_3)/(1 - r_2)$ and $y = 1$. In this case, these equilibria are precisely those combinations of strategies satisfying

$$s = 1; \quad u = 1 - t, \ t \geq \frac{1 - r_3}{x(1 - r_2)}$$

Payoffs at any equilibrium of type I are $(r_3, 1)$.

Analogously, we can identify all equilibria with $s = 0$. First, from (2),

$$E_C(u,t;0) = [1 - u - t(1 - x)]c_2$$

so Column maximizes his expected payoff by picking $u = 0$, $t = 0$, or, if

$x = 1$, $u = 0$, t arbitrary. If $x < 1$, then by (1),

$$E_R(s;0,0) = (1 - s)r_2$$

so $s = 0$ is indeed a best response for Row. If $x = 1$, then (1) shows that

$$E_R(s;0,t) = r_2 + s(tyr_3 - r_2)$$

so $s = 0$ maximizes Row's expected payoff iff

$$t \leq \frac{r_2}{yr_3}$$

(This inequality holds for any choice of t if $y \leq r_2/r_3$; if $y > r_2/r_3$, some values of t are excluded.) Because the above necessary conditions can also be shown to be sufficient, we have

II *Noncooperative equilibria.* There are essentially two equilibria of this type. The strategy combination

$$s = 0; \qquad u = 0, \ t = 0$$

is type IIa. There are no other equilibria with $s = 0$ unless $x = 1$, in which case the equilibria with $s = 0$ are precisely those strategy combinations satisfying

$$s = 0; \qquad u = 0, \ t \leq \frac{r_2}{yr_3}$$

which is type IIb. At any type II equilibrium, payoffs are (r_2, c_2).

Having identified all equilibria with $s = 0$ or $s = 1$, we turn to the case $0 < s < 1$. Differentiation of (1) yields

$$\frac{\partial E_R}{\partial s} = -r_2 + u(r_2 + r_3 - 1) + t[yr_3 - (1 - x)(1 - r_2)] \tag{12}$$

Observe that $\partial E_R/\partial s = 0$ at any equilibrium with $0 < s < 1$. It follows that there are no such equilibria with $t = 0$, for if $t = 0$,

$$\frac{\partial E_R}{\partial s} = -r_2 + u(r_2 + r_3 - 1) \leq \max\{-r_2, \ r_3 - 1\} < 0$$

by (12). Now suppose that (see figure 8.2)

$$x(1-r_2)+yr_3<1$$

Then

$$yr_3-(1-x)(1-r_2)=yr_3+x(1-r_2)-(1-r_2)$$
$$<1-(1-r_2)=r_2$$

so that (12) yields

$$\frac{\partial E_R}{\partial s}<-r_2+ur_2+tr_2=-r_2+(u+t)r_2\leqq0$$

since $u+t\leqq1$. Again, no equilibrium with $0<s<1$ can exist.

Thus, to find all equilibria not of types I or II, we may assume that (see figure 8.2)

$$x(1-r_2)+yr_3\geqq1 \tag{13}$$

Further, we need only consider strategies for Column with $t>0$. First consider the case $u=0$, $t=1$. Substitution in (12) shows that

$$-r_2+[yr_3-(1-x)(1-r_2)]=0$$

which is equivalent to

$$x(1-r_2)+yr_3=1$$

is necessary. From (2),

$$E_C(u,t;s)=sc_3+(1-s)c_2+u[s(1-c_3)-(1-s)c_2]$$
$$+t[sy(1-c_3)-(1-s)(1-x)c_2]$$

which we write as

$$E_C(u,t;s)=H+uK+tL \tag{14}$$

with H, K, and L defined appropriately. By calculus, the choice $u=0$, $t=1$, maximizes (14) iff

$$L\geqq0, \qquad L\geqq K$$

The condition $L\geqq0$ is easily seen to be equivalent to

$$s\geqq s_t(x,y)=\frac{(1-x)c_2}{y(1-c_3)+(1-x)c_2}$$

and the condition $L\geqq K$ is equivalent to

$$s \leqq s_u(x,y) = \frac{xc_2}{(1-y)(1-c_3) + xc_2}$$

It is obvious that $0 \leqq s_l < 1$, with equality iff $x = 1$, and $0 < s_u \leqq 1$, with equality iff $y = 1$. Finally,

$$s_u - s_l = c_2(1-c_3)(x+y-1) > 0$$

since $x + y > x(1-r_2) + yr_3 = 1$.

It can be verified that these necessary conditions are also sufficient for an equilibrium with $u = 0$, $t = 1$. In summary, we have

III *Constant-detection equilibrium.* There is an equilibrium with $0 < s < 1$, $u = 0$, and $t = 1$ iff $x(1-r_2) + yr_3 = 1$. In this case, such equilibria are precisely those strategy combinations satisfying

$$s_l \leqq s \leqq s_u; \qquad t = 1, \, u = 0$$

Payoffs at any equilibrium of type III are

$$(yr_3, \, xc_2 + s[c_3 + y(1-c_2) - xc_2])$$

Next, under the assumption that (13) holds, we search for equilibria with $0 < t < 1$ and $u = 0$, and, of course, $0 < s < 1$. Substitution in (12) shows that

$$-r_2 + t[yr_3 - (1-x)(1-r_2)] = 0$$

which is equivalent to

$$t = t_l(x,y) = \frac{r_2}{yr_3 - (1-x)(1-r_2)}$$

[As noted above, (13) ensures that the denominator of this fraction is at least r_2 so that $0 < t_l \leqq 1$, with equality precisely when equality obtains in (13).]

Now consider (14). If the choice $0 < t < 1$ and $u = 0$ maximizes Column's expected payoff, it is necessary that $L = 0$ and $K \leqq 0$. As above, $L = 0$ iff $s = s_l(x,y)$ and, assuming this to be true, $K \leqq 0$ iff

$$-c_2 + s_l(1 - c_3 + c_2) \leqq 0$$

This relationship can be shown to be equivalent to

$$c_2(1-c_3)(1-x-y) \leqq 0$$

which is certainly true, since, as noted earlier, (13) ensures that $x + y > 1$.

Because the above necessary conditions are also sufficient, we have

IV *Never-accept (unconditionally) equilibrium.* There is an equilibrium with $0 < s < 1$, $u = 0$, and $0 < t < 1$ iff $x(1 - r_2) + yr_3 > 1$. In this case, such equilibria are precisely those strategy combinations that satisfy

$$s = s_l; \qquad u = 0, \, t = t_l$$

Payoffs are any equilibrium of type IV are

$$\frac{yr_2 r_3}{yr_3 - (1 - x)(1 - r_2)}, \qquad \frac{yc_2(1 - c_3) + (1 - x)c_2 c_3}{y(1 - c_3) + (1 - x)c_3}$$

Analogously, we now assume (13) and search for equilibria with $u = 1 - t$, $0 < t < 1$, and, of course, $0 < s < 1$. Substitution in (12) and simplification show that

$$t = t_u(x,y) = \frac{1 - r_3}{x(1 - r_2) - (1 - y)r_3}$$

is a necessary condition for such an equilibrium. [It is easy to verify that $0 < t_u \leqq 1$, with equality precisely when equality holds in (13).] The choice of $0 < t < 1$ and $u = 1 - t$ maximizes (14) only if $L = K$ and $L \geqq 0$. As observed earlier, $L = K$ iff $s = s_u(x,y)$. If $s = s_u$, $L \geqq 0$ iff

$$-(1 - x)c_2 + s_u[y(1 - c_3) + (1 - x)c_2] \geqq 0$$

which is equivalent to

$$c_2(1 - c_3)(x + y - 1) \geqq 0$$

The latter inequality is a consequence of (13).

Because these necessary conditions are also sufficient, we have

V *Never-challenge (unconditionally) equilibrium.* There is an equilibrium with $0 < s < 1$, $u = 1 - t$, and $0 < t < 1$ iff $x(1 - r_2) + yr_3 > 1$. In this case, such equilibria are precisely those strategy combinations that satisfy

$$s = s_u; \qquad u = 1 - t_u, \, t = t_u$$

Payoffs at any equilibrium of type V are

$$\left(\frac{x(1-r_2)r_3 - (1-y)r_3}{x(1-r_2) - (1-y)r_3}, \frac{xc_2}{(1-y)\,(1-c_3) + xc_3} \right)$$

We now show that we have identified all Nash equilibria of the Verification Game with detection. If there is an equilibrium not already described, it can exist only when (13) holds, and it must satisfy $0 < s < 1$, $0 < u < 1 - t$, and $0 < t < 1$. Consideration of (14) shows that a necessary condition for such an equilibrium is $L = 0$ and $K = 0$.

We have already noted that $L = 0$ iff $s = s_l$, and $L = K$ iff $s = s_u$. Since $s_u = s_l$ can be shown to be equivalent to

$$c_2(1 - c_3)\,(x + y - 1) = 0,$$

and since $x + y > 1$ is a consequence of (13), we conclude that no such equilibrium can exist. Therefore, the only equilibria of the Verification Game with detection are those given by I–V.

Comparisons

When $x(1 - r_2) + yr_3 < 1$, the Verification Game with detection is simple: Row's strategy $s = 0$ (noncompliance) is dominant, and there is a unique (except when $x = 1$) equilibrium (IIa) with payoffs (r_2, c_2). On the other hand, the situation is more complicated when $x(1 - r_2) + yr_3 > 1$, for even with $x < 1$ and $y < 1$ there are two distinct new equilibria, IV and V, as well as the possibility of inducement by Column. We next compare these latter possibilities.

Assume that $x(1 - r_2) + yr_3 > 1$, $x < 1$, and $y < 1$. First observe that $s_u > s_l$ by a calculation similar to the one above. Also $t_u > t_l$ iff $1 - r_2 - r_3 > 0$, and $t_u = t_l$ iff $1 - r_2 - r_3 = 0$. Finally, comparison of (9) with the definition of t_u shows that Column's optimal inducement strategy is to consult his detector (D) just slightly more often than at the type V equilibrium, where the proportion is t_u.

We now compare the players' payoffs at type IV and type V to show that V dominates IV and that both dominate IIa. For Column, $E_C(\text{IV}) \geqq c_2$, with equality iff $x = 1$, as is obvious. The inequality $E_C(\text{V}) > E_C(\text{IV})$ is equivalent to

$$xc_2[y(1 - c_3) + (1 - x)c_2]$$
$$> c_2[(1 - y)\,(1 - c_3) + xc_2][y(1 - c_3) + (1 - x)c_3] \quad (15)$$

which can be shown to hold iff

$$x[y(1 - c_2) + (1 - x)c_2] > (1 - y)[y(1 - c_3) + (1 - x)c_3]$$

Now $x > 1 - y$ since (13) holds, and

$$[y(1-c_2)+(1-x)c_2]-[y(1-c_3)+(1-x)c_3]-(c_3-c_2)(x+y-1)>0$$

again because of (13). This proves that (15) holds, and we have $E_C(V) > E_C(IV) > E_C(II) = c_2$.

We now prove analogous inequalities for Row's payoffs at these three equilibria. First, it is easy to show that $E_R(IV) \geq E_R(IIa) = r_2$, with equality iff $x = 1$. The inequality $E_R(V) > E_R(IV)$ can be shown to be equivalent to

$$Q(y) = m - ny + py^2 > 0$$

where $m = (1-x)[1-x(1-r_2)]$, $n = (1-x)(1+r_3) - xr_2$, and $p = r_3$. It is easy to verify that if $(1-r_3)/(1-r_2) < x < 1$,

$$Q\left(\frac{1-x+xr_2}{r_2}\right) = 0$$

The minimum of $Q(y)$ occurs at $y = n/2p$, and again it can be checked that

$$\frac{n}{2p} < \frac{1-x+xr_2}{r_3} \text{ for } \frac{1-r_3}{1-r_2} < x < 1$$

This completes the proof that

$$E_R(V) > E_R(IV) > E_R(II) \text{ for } x(1-r_2)+yr_3 > 1, \ x<1, \ y<1$$

Therefore, of the three equilibria that exist in the region $x(1-r_2) + yr_3 > 1$, $x < 1$, and $y < 1$, V is strictly preferred by both players. It is immediately apparent that $E_R(V) < r_3$, and it can be shown that

$$E_C(V) \geq c_3 \quad \text{iff} \quad xc_2 \geq (1-y)c_3$$

We now compare the payoffs at the dominant equilibrium, V, with the payoffs under optimal inducement by Column. We have already noted that Column's strategies are only marginally different in these two situations. Comparison of $E_R(V)$ with (11) shows that Row receives slightly less under optimal inducement by Column than at V. [This also proves that $E_R(V) < r_3$.] Using (10), it can be shown that Column's expected payoff under optimal inducement, E_C^*, exceeds $E_C(V)$ exactly when

$$x(1 - r_2 - c_2 + c_2 r_3) + y(1 - c_3 + c_3 r_3) > 1 - c_3 + c_3 r_3 \qquad (16)$$

It is easy to verify that the line defined by equality in (16) lies below and to the left of $x(1 - r_2) + yr_3 = 1$ (see figure 8.2). Therefore, the region where $E_C^* > E_C(V)$ is the entire triangle defined by $x(1 - r_2) + yr_3 > 1$, $x < 1$, and $y < 1$, which is the shaded area shown in figure 8.2. Optimal inducement by Column thus increases Column's expected payoff significantly over equilibrium V. Also, recall that $E_C^* > c_3$ whenever inducement by Column is possible.

NOTES

This chapter is drawn from Steven J. Brams and D. Marc Kilgour, Verification and stability: a game-theoretic analysis, in *Arms and Artificial Intelligence: Weapons and Arms Control Applications of Advanced Computing*, Allan M. Din (ed.) (Oxford, UK: Oxford University Press, 1987), pp. 193–213. Reprinted with permission of the Stockholm International Peace Research Institute.

1   Recent books on verification include Scribner, Ralston and Metz (1985); Krass (1985); Rowell (1986); and Tsipis, Hafemeister, and Janeway (1986). Recently, verification may have lost its primacy as a divisive issue since Soviet leaders have agreed to certain kinds of on-site inspections. See Gwertzman (1986) and Tsipis (1987).

2   See Maschler (1966, 1967) and Rapoport (1966, pp. 158–185).

3   This game was originally suggested to Brams by Danny Kleinman in 1983; we are pleased to acknowledge its source.

4   See also Brams (1985b, ch. 4). For extensions of this analysis to other games, see Wittman (1985).

5   It can also be circumvented if Row can make a self-binding commitment to choose C, though it is hard to see how such a commitment in the Verification Game could be made credible to Column. See Maoz and Felsenthal (1987).

6   The probabilities of error are $1 - x$ and $1 - y$; in statistics, an error of the latter kind (incorrectly signaling a violation) is a type 1 error, and an error of the former kind (incorrectly signaling compliance) is a type 2 error.

7   We have ignored a fourth pure strategy for Column: consulting his detector and acting as if it is incorrect by choosing A if it signals a violation and $\overline{A}$ otherwise. Unless x and y are small, this fourth strategy would simply complicate the analysis without adding any significant features to the game. In fact, our analysis will focus on relatively large values of x and y.

8   Lynn (1985) gives details on bilateral and multilateral agreements and assesses compliance with them. See also Nincic (1986).

9   See Brams and Davis (1987) and Brams (1985b, ch. 4).

9

National Security and War

In this brief and final chapter, we take a sharply different tack. Putting aside our game-theoretic analysis for the moment, we begin by discussing the probability of nuclear war, illustrate the calculation of this probability with some hypothetical figures, and then state more general results. Our main purpose, however, is to show how our game-theoretic analysis bears on the *prevention* of nuclear and other wars, so we conclude by summarizing our principal findings bearing on this question in section 9.2.

9.1 THE PROBABILITY OF NUCLEAR WAR[1]

At the height of the Cuban missile crisis in October 1962, John F. Kennedy estimated the chances of a major war between the United States and the Soviet Union to be somewhere between one-third and one-half.[2] There is no indication that President Kennedy had any firm basis for making such an estimate, much less a model for deriving this prediction, but this intuition about the danger of the superpower confrontation possibly escalating to nuclear conflict was probably as good as anybody else's.

Other world leaders have frequently made predictions about the probability of nuclear war, as have military analysts, but these predictions have generally not been rooted in any systematic approach. Indeed, it is hard to see what kind of technique could be used to forecast reliably a unique event that involves two or more nuclear powers. The only prior wartime use of nuclear weapons was by the United States against Japan in 1945, and in that case only the United States had nuclear weapons. Moreover, many other circumstances have changed drastically since 1945, making extrapolations from this first wartime use dubious.

Most analysts view a "bolt from the blue"—a massive and unexpected first strike of one superpower against the other—as exceedingly unlikely.[3] Instead, the main danger is seen as a steady erosion of trust, which in an extreme crisis could precipitate the first use of nuclear weapons, particularly if the initiator faced a desperate situation and believed that only nuclear weapons might provide an escape from certain defeat. Alternatively, there is always some small but finite probability that nuclear weapons will be used accidentally because of potential failures in C^3I.[4]

Whatever the reason for the outbreak of nuclear war, we assume that there is some positive probability that it will occur in, say, the next year. If, for example, this probability is 1 percent per year into the indefinite future, then the probability is greater than .5 that nuclear war will occur within 69 years. This value is the minimum integer n satisfying the inequality

$$.01 + (1 - .01)(.01) + (1 - .01)^2(.01) + \ldots + (1 - .01)^{n-1}(.01) > .5$$

in which the terms on the left side are the (constant) conditional probabilities of nuclear war in years 1, 2, . . ., n, given that war did not occur in any preceding year, times the probability of no earlier war. It is not difficult to show that as the number of years goes to infinity, the probability of eventual nuclear war approaches 1, no matter how small (but positive and constant) the probability of war per year may be.

More conveniently for purposes of calculation, the above inequality is equivalent to

$$1 - (.99)^n > .5$$

where the left side can be interpreted as 1 minus the probability that nuclear war will not occur at some time within n years. Simplifying and taking logarithms gives

$$n > \frac{\log (.5)}{\log (.99)} = 68.97$$

confirming that the smallest integer solution is $n = 69$.

This is not a happy picture as one peers into the future. Indeed, it may be a future we want to shrink from, as Iklé somberly pointed out: "We all turn away . . . from the thought that nuclear war may be as inescapable as death, and may end our lives and our society within this generation or the next."[5]

More hopefully, suppose that the probability of nuclear war in each subsequent year can be decreased by, say, 20 percent per year. Then Garwin (1985, p. 39) observed that the cumulative probability of nuclear war—over an eternity—would be dramatically reduced to less than 5 percent. This result flows from a simple calculation, where the terms in the sum on the left side of the equation below are upper bounds for the unconditional probabilities of nuclear war in years 1, 2, . . . :

$$.01 + (.8)(.01) + (.8)^2(.01) + \ldots = (.01)\left(\frac{1}{1 - .8}\right) = .05$$

The evaluation of the expression follows from the identity

$$1 + x + x^2 + x^3 + \ldots = \sum_{i=0}^{\infty} x^i = \frac{1}{1 - x}$$

where $|x| < 1$, for the sum of an infinite geometric series. Our calculation gives an upper bound (the exact value is approximately 0.0489) for the *eventual* probability of nuclear war—that is, into the indefinite future—which is the sum of the above infinite series.

A theoretical analysis of the probability of nuclear war that generalizes this calculation demonstrates the following propositions:

1 Whatever the starting probability (1 percent in the above example), a *constant* reduction factor (20 percent in the above example) ensures that the eventual probability of nuclear war is always strictly less than 1.

2 If the reduction factor is not constant but decreases at a constant rate (say, the 20 percent factor itself decreases by 20 percent per year: 20 percent, 16 percent, and so on), the eventual probability of nuclear war is always 1.

With respect to the first proposition, trade-offs between the starting probability and the reduction factor are of some interest. The latter factor, if constant, turns out to be increasingly important as the starting probability falls. Nevertheless, the starting probability can have a substantial immediate effect. Thus, if in Garwin's example it rose from 1 percent to 5 percent, and the 20 percent reduction factor remained the same, the eventual probability of nuclear war would rise to about 18 percent.

Now assume that the starting probability remains constant at 1 percent but the reduction factor drops from 20 percent to 5 percent. Then the eventual probability of nuclear war would rise from about 5 percent to about 22 percent. This probability jumps to about 64 percent when the starting probability is 5 percent and the reduction factor is also 5 percent.

One major step in averting nuclear war is to make sure that its eventual probability does not go to 1. We will always be successful in this task, if, no matter what the probability of nuclear conflict in the next year is, we can succeed in reducing it by at least a constant factor every year thereafter. Of course, if we can reduce the probability of war not just at a constant rate but at an increasing rate, then the eventual probability is diminished even farther below 1 than a constant rate of decrease would imply.

More ominous, nuclear war becomes a certainty eventually, whatever the starting probability, if the annual reduction is not constant but decreases at a constant rate. That is, if the probability of nuclear war in each subsequent year, even though reduced, is reduced slowly enough, then war will occur in the proverbial long run. (War may still occur if the reduction factor is constant, but it is not a *certainty* unless this factor is decreasing.) Thus, the economist's usual assumption of marginally decreasing returns, if applicable to the year-to-year reduction proportion, is not a good omen: When the probability of war decreases at a decreasing rate, war eventually becomes inevitable.

Our feeling is that it is unlikely, short of either the abolition of nuclear weapons or perfect Star Wars defenses—whose likelihood seems negligible—that the annual probability of nuclear war will eventually go to zero. We think that, probably beginning after the Cuban missile crisis, the probability of nuclear war, at least between the superpowers, has decreased, perhaps markedly. Speculatively, it may now be not 5 or 1 percent in the next year but considerably less.[6]

Thus, even if eventual nuclear war is a certainty, it may be a long time in coming. In fact, though the dangers of nuclear war probably wax and wane, the world is likely much safer than it was a generation ago when the superpowers had not yet acquired significant second-strike retaliatory capabilities and nuclear deterrence was decidedly less stable than today.

Nonetheless, the memory of the Cuban missile crisis has receded, and it may take another searing experience near the nuclear precipice to rejuvenate efforts to avoid nuclear confrontations and slow down the nuclear arms race. The core meltdown at Chernobyl in April 1986 reawakened fears of a nuclear disaster and may have given new momentum to the development of additional safeguards against the accidental or deliberate use of nuclear weapons—as well as the handling of fissionable material—just as the Cuban missile crisis led to the hot line between the superpowers a generation earlier.

Our calculations show how both the starting probability and the reduction factor, especially the former when the starting probability is low, act together to diminish the probability of nuclear war. A crisis or disaster—if survived—probably lowers both, but ultimately its sobering effects may prove ephemeral as memories fade.

The key, we think, is to try to set in motion—especially after a crisis or disaster when citizens and governments are aroused and apprehensive—processes that lead in stages to continued reductions in the probability of war. If these agreed-upon reductions are costly for the superpowers to back out of or renege on and hence stay in place reasonably well, then the cumulative probability of nuclear war, even over long (if not infinite) stretches, may be kept significantly below 1.

Moreover, if we survive into the foreseeable future, each year that we do so will lead to a new lease on life that extends farther and farther ahead. To be sure, in an infinite future we will all perish if the possibility of nuclear war is never eliminated altogether, but lengthening that future by steady and ineluctable reductions in the probability of nuclear war, while still allowing for some untoward developments along the way, could help immensely.

9.2 PREVENTING WAR

Our game-theoretic analysis in the preceding chapters may, from a normative perspective, perhaps best be viewed as suggesting strategies for minimizing the effect of adverse developments or, better yet, arresting their occurrence entirely. For instance, we found generally that retaliatory threats are essential to stabilizing the cooperative outcomes in all the games studied. Because a departure from these outcomes to score a "win" could elicit a response that is more damaging to the attacker than staying at the status quo, "winning" is correspondingly downgraded and may even, especially in the nuclear case, be rendered meaningless.

This may all sound obvious, if not banal; but we would caution the reader that, like most general findings, it only sets the stage. Game-theoretic models enable one to fill in details that may go well beyond what can be inferred informally from a general but shallow understanding of a strategic situation.

Before highlighting some of our more specific findings, more should be said about the role of threats in stabilizing a fragile peace. To many people, the very idea of threatening another party is inimical to generating good will and trust.[7] Moreover, threats may not only frighten but also provoke a party. We would agree to some extent, depending on how the threats are made as well as on the degree of amicability or animosity that surrounds an existing relationship between two parties. Unquestionably, intimidating behavior can raise tensions (as well as hackles) and erode trust quickly.

On the other hand, insofar as a relationship can be modeled broadly by the Conflict Game (chapter 1)—in which each party is tempted to defect from an unstable compromise outcome—then we see no alternative to using threats to inhibit the breaking of this compromise. International relations is not a tea party, and much as we would like to see everybody on their best behavior, rules that punish flagrant aggression seem essential to prevent coercive and bellicose behavior.

It is perhaps the pristine simplicity and obviousness of this, our central finding, that explains why it has sometimes been forgotten as a policy prescription by world leaders. There may be another explanation, connected with the all-too-human fixation on "winning" games—and, in the United States, on not "losing" wars. Because national security games are almost always variable-sum games, the goal of winning must generally be traded off against the goal of stabilizing cooperation by deterring aggression and preserving some kind of order.

If winning is not in the cards, however, it does not follow that splendid relations will spring forth effortlessly and flourish without risk. Thus, our search in the games we have analyzed has been for conditions, generally involving the possibility of retribution, that foster good behavior by turning Pareto-superior compromise outcomes into Nash equilibria.

The formal incorporation of threats into the rules of a game is a relatively new development. The founding work on game theory, von Neumann and Morgenstern's *Theory of Games and Economic Behavior*, was a stunning intellectual achievement, yet it does not even have the word "threat" in its index.[8] A generation later, Schelling, in his pioneering *The Strategy of Conflict*, showed informally that threats can play an absolutely fundamental role in the structuring of conflict, but he did not develop any formal theory of threats.[9] It is perhaps this intellectual lacuna in the mathematical theory, as much as anything else, that accounts for the lack of formal models explicitly incorporating threats into the study of national security.

This is not to say that the national security literature is lacking in casual advice and haphazard prescriptions espousing when and how to threaten adversaries—or, more euphemistically, promote "reciprocity." Unfortunately, most of this literature is informally derived from "lessons" learned in recent history or from some rather loose and speculative theorizing.

Our own effort to be more formal in our derivations and more systematic in our theorizing has the virtue, we believe, that our conclusions are more accessible to careful review and revision. Because they follow deductively from our assumptions—essentially, the rules of the games we posit—they can readily be altered when found wanting. New rules can be formulated and new conclusions derived, generating a new set of theoretical consequences, whose empirical adequacy can be tested. This interactive process can be repeated indefinitely.

If the standard view of scientific theorizing (and progress) is valid, this process should be self-correcting. Indeed, to facilitate learning in the national security field, we would suggest the following sequence: First, national security analysts might begin by trying to formulate the rules of different games that seem to capture salient strategic features they wish to model. Then, perhaps with the aid of formal theorists, they could ascertain what outcomes are stable in these games and the strategies required to support this stability. These could then be checked against actual behavior, and the process repeated.

We are not suggesting, by the way, that stability is the be-all and end-all of a satisfactory international order. Also, as we indicated earlier (see, particularly, references on p. 19 and in note 7, p. 125), we are well aware that Nash's concept of equilibrium is just one stability concept among several. In addition, should there be no stability, game theory offers some mathematical structures for exploring dynamic processes that may not have any single resting place but instead may stimulate a kind of shifting order, constantly in motion. Finding this order, and characterizing it game-theoretically, will pose new intellectual challenges for formal theorists.

Perhaps this brief digression on theorizing about national security sets the stage for summarizing our main results. Beyond our major finding that the threat of retaliation is crucial to stabilizing cooperation in the games studied, we showed the following in chapters 1–8:

1 To the extent that the Conflict Game provides a generic representation of conflict between two parties (states or alliances) in international relations, different strategies may lead to cooperation, but only one (incorporating some form of tit-for-tat) makes it stable. The threat of retaliation may require precommitments to ensure its credibility, but the threat need only be probabilistic, not carried out with certainty.

2 In the Deescalation Game (based on Prisoners' Dilemma), the noncooperative outcome is always a Nash equilibrium, as it is in Prisoners' Dilemma, but probabilistic threats can render the cooperative outcome, unstable in the classical 2×2 game, stable as well. Moreover, unlike Prisoners' Dilemma, either player can initiate a move from the Pareto-inferior escalation equilibrium to the Pareto-superior deescalation equilibrium. The initial step is costless and induces subsequent rational moves that benefit both players, eventually leading to the deescalation equilibrium. The superpower arms race seems well modeled by the Deescalation Game, which suggests, prospectively, possible ways that arms control agreements can be reached.

3 In the Deterrence Game (based on Chicken), the Nash equilibria essentially duplicate those in Chicken, except for a deterrence equilibrium at which the players never preempt but are always prepared to retaliate with a probability above a calculable

threshold. This equilibrium is Pareto-superior, dynamically stable, and—when supported by "robust threats"—as invulnerable as possible to misperceptions or miscalculations by the players. The precommitments of the superpowers to retaliate if attacked, because of the uncertainties inherent in C^3I, are consistent with probabilistic threats in the Deterrence Game. These pre-commitments render the threat of retaliation credible even if retaliation leads to a worse outcome—perhaps a nuclear holocaust—than a side would suffer from absorbing a limited first strike and not retaliating.

4 In two games analyzed in chapter 4, an exogenous probability of winding down is postulated whereby the players can escape either the mutually worst outcome in the Deterrence Game (which becomes the Winding-Down Game) or the mutually next-worst outcome in the Deescalation Game (which becomes the Arms Reduction Game). In the former game, winding down is consistent, but only up to a point, with making mutual cooperation a dominant-strategy Nash equilibrium through precommitments to probabilistic threats, whereas in the Arms Reduction Game winding down is always consistent with making mutual deescalation a Nash, but not a dominant-strategy, equilibrium. In both games, as the probability of winding down increases, the threat of retaliation must also increase—and at an increasing rate—to preserve the stability of the cooperative outcome (mutual deterrence or mutual deescalation). These trade-offs suggest that there is no "free lunch": Escape mechanisms (including Star Wars), if available should deterrence fail or an arms race occur, may themselves encourage defections from cooperation.

5 In the Star Wars Game, the defensive capabilities that each side possesses impose additional constraints on the choices of the players in the Deterrence Game. The Nash equilibria derived in this game, and illustrated for three different scenarios involving various postulated relationships between the first-strike and second-strike defenses of the players, often overlap. They demonstrate that, unlike in the Deterrence Game, *mutual* preemption may be an equilibrium, underscoring the problem—particularly if defensive capabilities are unbalanced—that deterrence may be subverted by the development of Star Wars. The analysis suggests that the superpowers, in deploying Star Wars systems, must adhere to a tricky time path to avoid preemption and preserve crisis stability.

6 The Threat Game, also based on Chicken, permits the players to tailor their threats to the (initial) level of preemption of their opponents. There is a deterrence equilibrium in this game that renders any level of preemption by an opponent costly. The strategies that support this equilibrium may entail more-than-

proportionate threats of retaliation against low levels of preemption but less-than-proportionate threats against high levels, with the threat level always decreasing relative to the preemption level. Proportional retaliation (tit-for-tat) is a special case, and, in general, not optimal for deterring preemption at minimal cost.

7 In a crisis, the players in the Threat Game are assumed to have escalated their conflict and to desire to stabilize it before it explodes. This they can always do, but sometimes only by threatening their opponents more severely than before the crisis erupted, thereby heating up an already tense situation. Crisis stabilization is aided by being close to the full-cooperation position, although, paradoxically, both players may benefit from having created a crisis that only escalating threats may resolve. Also, if one player is much more cooperative in the crisis, or the players value the cooperative outcome relatively highly, crisis stabilization is facilitated.

8 The Verification Game is an asymmetrical game between an inspector and an inspectee. The inspectee most wants his claimed compliance with an arms control treaty accepted, whereas the inspector most desires actual compliance but can only imperfectly detect the actions of the inspectee. Given that the inspector's detection probabilities are above a certain threshold, inducement by the inspector is possible, and the existence of a Pareto-superior Nash equilibrium is guaranteed. However, weapons such as cruise missiles that are difficult to detect and verify may place the superpowers below this threshold, precluding all but a non-cooperative and Pareto-inferior Nash equilibrium, at which the inspectee always cheats and the inspector never believes the inspectee's claimed compliance.

This summary, of course, strips away much of the fine detail of the models, just as the game-theoretic models themselves abstract much from the richness of real-life national security games. The models offer a start, nevertheless, to the rigorous and in-depth analysis of important strategic features of national security issues.

Deriving general conclusions about rational behavior in arms races, crises, and the like will, we hope, help us avoid them or, that failing, minimize their most pernicious and terrifying consequences. It should also help us to diminish the cumulative probability of nuclear war, whose prevention over time, as we have seen, requires constant reductions in its probability each year.

NOTES

1 This section is drawn in part from Rudolf Avenhaus, Steven J. Brams, John Fichtner, and D. Marc Kilgour, The probability of nuclear war (mimeographed, 1986), which contains proofs of the general results briefly described later.

2 Sorensen (1965, p. 705). However, other participants in the crisis thought the probability of nuclear war was considerably lower; see Lebow, (1987, p. 15).

3 See, for example, Bracken (1983, p. 73).

4 Vulnerabilities in C^3I are discussed in Blair (1985); Ford (1985); and Bracken (1983). The relationship of "security reliability" to the probability of nuclear war is analyzed in Cioffi-Revilla (1987).

5 Iklé (1973, p. 267). See also Quester (1986) and Nye (1987).

6 This tends to be the assessment of specialists. See Nye, Allison, and Carnesale (1985, p. 207), who add, "Not much is proven by this finding. In this realm there are 'specialists' but not 'experts.'" Bracken (1985, p. 50) offers a somewhat higher range of estimates of the probability of *accidental* nuclear war (averaging about 10 percent over the next 10 years).

7 Boulding (1978, p. 157), for example, contends that threats may be pathological in the sense of crippling the learning process—"we learn not to learn."

8 von Neumann and Morgenstern (1953) is the third edition. This work was first published in 1944.

9 Schelling (1960). A game-theoretic analysis of threats, based on Schelling (1966), is given in Brams and Hessel (1984).

Glossary

This glossary contains definitions of game-theoretic and related terms used in this book. Specific games and their Nash equilibria that have been developed in detail in the text are not included. As in the text, an attempt has been made to define these terms in relatively nontechnical language.

Certain equivalent A certain equivalent is a deterministic action (for example, a lower-level retaliatory action) whose value or worth to an actor is the same as the expected value of an action (e.g., full-scale retaliation) selected not for certain but probabilistically.

Complete information A game is one of complete information if each player knows the rules of play of the game and the preferences, or payoffs, of every player over all possible outcomes.

Conditional probability A conditional probability is a probability that is contingent on the occurrence of a prior event.

Constant-sum game A constant-sum game is a game in which the payoffs to the players at every outcome sum to some constant, so if the game has two players, what one player gains the other loses.

Crisis In the Threat Game, a crisis is an event, or series of events, that changes the preplay position of at least one player to one of less-than-full cooperation.

Crisis stability Crisis stability occurs when no player in a crisis has an incentive to escalate a conflict further.

Crossover point In the Threat Game, a player's crossover point is the exact level of provocation at which the retaliation-to-provocation (RP) ratio is 1—that is, where tit-for-tat is just sufficient for deterrence.

Decision theory Decision theory is a mathematical theory for making optimal choices, based on the assumption that the outcome does not depend on the choices of other decision makers but rather on states of nature that arise exogenously according to a probability distribution and on the decision maker's own choices.

Deterrence Deterrence is a policy of threatening retaliation against preemption by an opponent to deter that preemption in the first place.

Dominant strategy A dominant strategy is a strategy that leads to outcomes at least as good as those of any other strategy in all possible contingencies, including a strictly better outcome in at least one contingency. A strictly dominant strategy is a dominant strategy that leads to a better outcome in every contingency.

Dominated strategy A dominated strategy is a strategy that leads to outcomes no better than those given by some other strategy in all possible contingencies that may arise, including a strictly worse outcome in some contingency.

Dynamic stability An outcome is dynamically stable if, given that one player departs from it, the other player(s) will have no incentive also to depart.

Expected payoff An expected payoff is the sum of the payoff a player receives from each outcome, multiplied by its probability of occurrence, for all possible outcomes that may arise.

Final position In the Threat Game, the final position is the position of the players on the game board after the player who was less preemptive in the first stage has retaliated in the second stage.

Game A game is the sum total of the rules of play that describe it.

Game board In the Threat Game, the game board is the unit square on which the players jointly determine different positions (preplay, initial, and final).

Game of partial conflict A game of partial conflict (variable-sum game) is a game in which the players' preferences are not diametrically opposed.

Game theory Game theory is a mathematical theory of rational strategy selection used to analyze optimal choices in interdependent decision situations, wherein the outcome depends on the choices of two or more actors or players, and each player has his own preferences over all possible outcomes.

Inducement strategy In the Verification Game, an inducement strategy is one in which an inducer offers an incentive to an inducee to choose a particular strategy in order to maximize his (the inducee's) expected payoff in a way that favors the inducer.

Initial position In the Threat Game, the initial position is the position of the players on the game board after they have chosen their preemption strategies—that is, after the first stage.

Lexicographic decision rule A lexicographic decision rule is a system for describing a decision maker's preferences over outcomes that uses a most important criterion ("primary goal"), then a next most important criterion ("secondary goal"), and so on.

Maximin strategy A maximin strategy is a strategy that maximizes the minimum payoff that a player can receive.

Minimax strategy A minimax strategy is a strategy that minimizes the maximum payoff that an opponent can receive.

Mixed strategy A mixed strategy is a strategy that involves a random selection from two or more pure strategies, according to a particular probability distribution.

Nash equilibrium A Nash equilibrium is an outcome from which no player would have an incentive to depart unilaterally because he would immediately do worse, or at least not better, if he did.

National security National security refers to those circumstances or events that directly affect the safety or integrity of states in their relations with other states.

Noncooperative game A noncooperative game is a game in which the players cannot make binding or enforceable agreements.

Normal form A two-person game is represented in normal form when it is described by an outcome or payoff matrix in which players are assumed to choose their strategies (rows or columns) independently. The possible outcomes of the game correspond to the cells of the matrix.

Ordinal game An ordinal game is a game in which each player can rank the outcomes but not necessarily assign payoffs or cardinal utilities to them.

Outcome/payoff matrix An outcome/payoff matrix is a rectangular array, or matrix, whose entries indicate the outcomes or payoffs to each player resulting from each of their possible strategy choices.

Pareto-inferior/superior outcome An outcome is Pareto-inferior if there exists another outcome that is better for some players and not worse for any other player. If there is no other outcome with this property, then the outcome in question is Pareto-superior.

Payoff A payoff is a measure of the value that a player attaches to an outcome in a game. Usually payoffs are taken to be cardinal (von Neumann–Morgenstern) utilities.

Perfect equilibrium A perfect equilibrium is a Nash equilibrium that is supported by threats that it would be rational for the threatener to carry out.

Perfect information A game is one of perfect information if each player, at each stage of play, knows the strategy choices of every other player up to that point.

Points of threat escalation In the Threat Game, points of threat escalation are the points on the game board whose stabilization requires retaliatory threats more severe than those described by the players' basic threat lines, which support the cooperative equilibrium of the preplay position.

Preference The preference of a player is his ranking of outcomes from best to worst.

Preplay position In the Threat Game, the preplay position is the position of the players on the game board at the start of play, before they choose their preemption and retaliation strategies.

Probabilistic threat A probabilistic threat is a threat that is carried out with a probability less than 1.

Pure strategy A pure strategy is a single specific strategy.

Rational player A rational player is one who makes choices in order to attain better outcomes, according to his preferences or goals, in light of the presumed rational choices of other players in a game.

Retaliation-to-provocation (RP) ratio In the Threat Game, the RP ratio is the ratio of a player's minimal level of retaliation necessary to deter each level of provocation (or preemption) by an opponent.

Robust threats (deterrence strategies) Robust threats ensure that each player will suffer the same loss in switching from cooperation to noncooperation, whatever strategy his opponent chooses.

Rules of play The rules of play of a game describe the choices of the players at each stage of play, and how the outcome depends on these choices.

Security level The security level of a player is the best outcome or payoff he can ensure for himself, whatever strategies the other players choose.

State of nature A state of nature is a situation or set of circumstances that arises by chance in the world and not as a result of any decision maker's choices.

Strategy A strategy is a complete plan that specifies the exact course of action a player will follow no matter what contingency arises.

Symmetric game A symmetric game is one that is strategically the same for all players. In a two-person, normal-form symmetric game with ordinal payoffs, the players' ranks of the outcomes along the main diagonal are the same, whereas their ranks of the off-diagonal outcomes are mirror images of each other.

Threat line In the Threat Game, a threat line is a line (or curve) indicating the minimal level of retaliation necessary to deter each level of provocation by an opponent from any preplay position. When the preplay position is full cooperation, the threat line is called the *basic threat line*.

Tit-for-tat Tit-for-tat is the strategy of cooperating initially and subsequently choosing a level of retaliation exactly equal to the level of provocation of an opponent.

Undominated strategy An undominated strategy is a strategy that is not dominated by any other strategy.

Variable retaliation Variable retaliation is retaliation that depends on (is a nonconstant function of) the level of preemption of an opponent.

Variable-sum game A variable-sum game is a game in which the sum of the payoffs to the players at different outcomes is not constant but variable, so the players may gain or lose simultaneously at different outcomes.

Bibliography

Abt, Clark C. 1985. *A Strategy for Terminating a Nuclear War.* Boulder, Co.: Westview.

Altfeld, Michael F. 1985. Uncertainty as a deterrence strategy: a critical assessment. *Comparative Strategy* 5, no. 1: 2–26.

Avenhaus, Rudolf. 1986. *Safeguard Systems Analysis.* New York: Plenum.

——, Brams, Steven J., John Fichtner, and D. Marc Kilgour. 1986. The probability of nuclear war. Mimeographed.

Axelrod, Robert. 1984. *The Evolution of Cooperation.* New York: Basic.

——Personal communication to Steven J. Brams, February 26, 1985.

Ball, Desmond. 1981. Can nuclear war be controlled? Adelphi Paper No. 169. London: International Institute for Strategic Studies.

Ball, Desmond and Jeffrey Richelson (eds.) 1986. *Strategic Nuclear Targeting.* Ithaca, NY: Cornell University Press.

Bartholdi, John J., III, C. Allen Butler, and Michael A. Trick. 1986. More on the evolution of cooperation. *Journal of Conflict Resolution* 30, no. 1 (March): 141–70.

Behr, Roy L. 1981. Nice guys finish last—sometimes. *Journal of Conflict Resolution* 25, no. 2 (June): 289–300.

Bendor, Jonathan. 1987. In good times and bad: reciprocity in an uncertain world. *American Journal of Political Science* 31, no. 3 (August): 531–558.

Bennett, P. G. 1987. Beyond game theory—where? In *Analyzing Conflict and Its Resolution: Some Mathematical Contributions*, P.G. Bennett (ed.) Oxford, UK: Clarendon, pp. 43–69.

Berger, Adolf. 1953. *Encyclopedic Dictionary of Roman Law, Transactions of the American Philosophical Society, New Series* 43, pt. 2. Philadelphia: American Philosophical Society.

Bert, Rolf, and Adam-Daniel Rotfeld. 1986. *Building Security in Europe: Confidence-Building Measures and the CSCE [Conference on Security and Cooperation in Europe].* East-West Monograph Series, No. 2, Allen Lynch (ed.) New York: Institute for East-West Security Studies.

Betts, Richard K. 1987. *Nuclear Blackmail and Nuclear Balance.* Washington, DC: Brookings.

Blair, Bruce G. 1985. *Strategic Command and Control: Redefining the Nuclear Threat.* Washington, DC: Brookings.

Borawski, John (ed.) 1986. *Avoiding War in the Nuclear Age: Confidence-Building Measures for Crisis Stability.* Boulder, CO: Westview.

Boulding, Kenneth E. 1978. *Ecodynamics: A New Theory of Societal Evolution.* Beverly Hills, CA: Sage.

Bowman, Robert M. 1986. *Star Wars: A Defense Expert's Case Against the Strategic Defense Initiative*. Los Angeles: Jeremy P. Tarcher.

Bracken, Jerome. 1986. Stability, SDI, air defense and deep cuts. In *Modelling and Analysis in Arms Control*, Rudolf Avenhaus, Reiner K. Huber, and John D. Kettelle (eds.) Berlin: Springer-Verlag, pp. 183–214.

Bracken, Paul. 1985. Accidental nuclear war. In *Hawks, Doves, and Owls: An Agenda for Avoiding Nuclear War*, Graham Allison, Albert Carnesale, and Joseph S. Nye, Jr. (eds.) New York: Norton, pp. 25–53.

——. 1983. *The Command and Control of Nuclear Forces*. New Haven, CT: Yale University Press.

Brams, Steven J. 1980. *Biblical Games: A Strategic Analysis of Stories in the Old Testament*. Cambridge, MA: MIT Press.

——. 1983. *Superior Beings: If They Exist, How Would We Know? Game-Theoretic Implications of Omniscience, Omnipotence, Immortality, and Incomprehensibility*. New York: Springer-Verlag.

——. 1984. Letter in response to Zraket, "Strategic command, control, communications, and intelligence." *Science* 226, no. 4676 (16 November): 782.

——. 1985a. *Rational Politics: Decisions, Games, and Strategy*. Washington, DC: CQ Press.

——. 1985b. *Superpower Games: Applying Game Theory to Superpower Conflict*. New Haven, CT: Yale University Press.

——. and Morton D. Davis. 1987. The verification problem in arms control: a game-theoretic analysis. In *Interaction and Communication in Global Politics*, Claudio Ciotti-Revilla, Richard L. Merritt, and Dina A. Zinnes (eds.) London: Sage, pp. 141–161.

——, ——, and Philip D. Straffin, Jr. 1979a. The geometry of the arms race. *International Studies Quarterly* 23, no. 4 (December): 567–88.

——, ——, and ——. 1979b. A reply to Dacey, "Detection and disarmament: a comment on 'The geometry of the arms race.' " *International Studies Quarterly* 23, no. 4 (December): 599–600.

——, ——, and ——. 1984. Comment on Wagner, "The theory of games and the problem of international cooperation." *American Political Science Review* 78, no. 2 (June): 495.

—— and Marek P. Hessel. 1984. Threat power in sequential games. *International Studies Quarterly* 28, no. 1 (March): 15–36.

—— and D. Marc Kilgour. 1985a. Optimal deterrence. *Social Philosophy & Policy* 3, no. 1 (Autumn): 118–35. Reprinted in *Nuclear Rights/Nuclear Wrongs*, Ellen Frankel Paul et al. (eds.) Oxford, UK: Basil Blackwell, 1986, pp. 118–35; and in *Naturalism and Rationality*, Newton Garver and Peter H. Hare (eds.) Buffalo, NY: Prometheus, 1986, pp. 241–62.

—— and ——. 1985b. The path to stable deterrence. In *Dynamic Models of International Conflict*, Urs Luterbacher and Michael D. Ward (eds.) Boulder, CO: Lynne Rienner, pp. 11–25. Reprinted in *Exploring the Stability of Deterrence*, Jacek Kugler and Frank C. Zagare (eds.) Boulder, CO: Lynne Rienner, 1987, pp. 107–22.

—— and ——. 1986a. Rational deescalation. In *Evolution, Games, and Learning: Models for Adaptation in Machines and Nature*, Doyne Farmer, Alan Lapedes, Norman Packard, and Burton Wendroff (eds.) Amsterdam: North-Holland, pp. 337–50. Reprinted from *Physica D*, vol. 22.

—— and ——. 1986b. Is nuclear deterrence rational? *PS* 19, no. 3 (Summer): 645–51.

—— and ——. 1986c. Notes on arms-control verification: a game-theoretic analysis. in *Modelling and Analysis in Arms Control*, Rudolf Avenhaus, Reiner K. Huber, and John D. Kettelle (eds.) Berlin: Springer-Verlag, pp. 409–19.

—— and ——. 1987a. Threat escalation and crisis stability: a game-theoretic analysis. *American Political Science Review* 81, no. 3 (September): 833–50.

—— and ——. 1987b. Winding down if preemption or escalation occurs: a game-theoretic analysis. *Journal of Conflict Resolution* 31, no. 4 (December).

—— and ——. 1987c. Verification and stability: a game-theoretic analysis. In *Arms and Artificial Intelligence: Weapon and Arms Control Applications of Advanced Computing*. Allan M. Din (ed.) Oxford, UK: Oxford University Press, pp. 193–213.

—— and ——. 1987d. Optimal threats. *Operations Research* 35, no. 4 (July–August).

—— and ——. 1987e. Is nuclear deterrence rational, and will Star Wars help? *Analyse & Kritik* 9 (October): 62–74.

—— and ——. 1988. Deterrence versus defense: a game-theoretic model of Star Wars. *International Studies Quarterly* 32, no. 1 (March).

—— and Donald Wittman. 1981. Nonmyopic equilibria in 2×2 games. *Conflict Management and Peace Science* 6, no. 1 (Fall): 39–62.

Brodie, Bernard (ed.) 1946. *The Absolute Weapon: Atomic Power and World Order*. New York: Harcourt, Brace.

Brzezinski, Zbigniew (ed.) 1986. *Promise or Peril: The Strategic Defense Initiative*. Washington, DC: Ethics and Public Policy Center.

Bueno de Mesquita, Bruce. 1981. *The War Trap*. New Haven, CT: Yale University Press.

——. 1985. The war trap revisited: a revised expected utility model. *American Political Science Review* 79, no. 1 (March): 156–73.

—— and David Lalman. 1986. Reason and war. *American Political Science Review* 80, no. 4 (September): 1113–29.

Bundy, McGeorge, et al. 1986. Back from the brink. *Atlantic Monthly* 258, no. 2 (August): 35–41.

Bunn, Matthew, and Kosta Tsipis. 1983. The uncertainties of a preemptive nuclear attack. *Scientific American* 249. no. 5 (November): 38–47.

Canavan, Gregory H. 1985. Simple discussion of the stability of strategic defense. Los Alamos, NM: Los Alamos National Laboratory. April.

Carnesale, Albert. 1987. Commentary on Charles L. Glaser. In *Strategic Defenses and Soviet–American Relations*. Samuel F. Wells, Jr., and Robert S. Litwak (eds.) Cambridge, MA: Ballinger, pp. 175–77.

Carter, Ashton B., John D. Steinbruner, and Charles A. Zraket (eds.) 1987. *Managing Nuclear Operations*. Washington, DC: Brookings.

Chrzanowski, Paul. 1985a. Crisis stability during a transition to a deterrence posture reliant on defenses. Livermore, CA: Lawrence Livermore National Laboratory, UCID 20590. October.

——. 1985b. Strategic defense and crisis stability. Livermore, CA: Lawrence Livermore National Laboratory, UCID 20699, December.

Cioffi-Revilla, Claudio. 1983. A probability model of credibility. *Journal of Conflict Resolution* 27, no. 1 (March): 73–108.

——. 1987. Crises, war, and security reliability. In *Interaction and Communication in Global Politics*, Claudio Cioffi-Revilla, Richard L. Merritt, and Dina A. Zinnes (eds.) London: Sage, pp. 49–66.

Dacey, Raymond. 1979. Detection and disarmament: a comment on "The geometry of the arms race." *International Studies Quarterly* 23, no. 4 (December): 589–98.

——. 1987. Ambiguous information and the manipulation of plays of the arms race game and the mutual deterrence game. In *Interaction and Communication in Global Politics*, Claudio Cioffi-Revilla, Richard L. Merritt, and Dina A. Zinnes (eds.) London: Sage, pp. 163–79.

—— and Norman Pendegraft. 1986. The optimality of tit-for-tat. College of Business and Economics, University of Idaho. Mimeographed.

Dallmeyer, Dorinda G. (ed.) 1986. *The Strategic Defense Initiative: New Perspectives on Deterrence*. Boulder, CO: Westview.

DeNardo, James. 1987. Are strategic defenses strategically defensible? Department of Politics, Princeton University. Mimeographed.

Drell, Sidney D., Philip J. Farley, and David Holloway (eds.) 1985. *The Reagan Strategic Defense Initiative: Technical, Political, and Arms Control Assessment*. Cambridge, MA: Ballinger.

Fichtner, J. 1986. On solution concepts for solving two person games which model the verification problem in arms control. In *Modelling and Analysis in Arms Control*, Rudolf Avenhaus, Reiner K. Huber, and John D. Kettelle (eds.) Berlin: Springer-Verlag, pp. 421–41.

Fishburn, Peter C. 1974. Lexicographic orders, decision rules and utilities: a survey. *Management Science* 20, no. 11 (July): 1442–71.

Ford, Daniel. 1985. *The Button: The Pentagon's Command and Control System*. New York: Simon and Schuster.

Fraser, Niall M., and Keith W. Hipel. 1984. *Conflict Analysis: Models and Resolutions*. New York: North-Holland.

Friedman, Thomas L. 1986. Israel and the bomb: megatons of ambiguity. *New York Times*, November 9, p. 23.

Garthoff, Raymond L. 1982. The role of nuclear weapons: Soviet perceptions. In *Nuclear Negotiations: Reassessing Arms Control Goals in U.S.–Soviet Relations*, Alan F. Neidle (ed.) Austin, TX: Lyndon B. Johnson School of Public Affairs, pp. 1–11.

Garwin, Richard. 1985. Quoted in "Forum: is there a way out?" *Harper's* 270, no. 1621 (June): 35–47.

Gauthier, David. 1984. Deterrence, maximization, and rationality. *Ethics* 94, no. 3 (April): 474–95.

George, Alexander. 1986. Problems of crisis management and crisis avoidance in US-Soviet relations. In *Studies of War and Peace*, Øyvind Østerud (ed.) Oslo, Norway: Norwegian University Press, pp. 202–26.

Glaser, Charles L. 1987. Managing the transition. In *Strategic Defenses and Soviet-American Relations*, Samuel F. Wells, Jr., and Robert S. Litwak (eds.) Cambridge, MA: Ballinger, pp. 141–61.

Goodin, Robert E. 1984. Itinerants, iterations, and something in-between. *British Journal of Political Science* 14, pt. 1 (January): 129–32.

Güth, Werner. 1985. An extensive game approach to model the nuclear deterrence debate. *Zeitschrift für die gesamte Staatswissenschaft/Journal of Institutional and Theoretical Economics* 14: 525–38.

——. 1986. Deterrence and incomplete information: the game theory approach. In *Modelling and Analysis in Arms Control*, Rudolf Avenhaus, Reiner K. Huber, and John D. Kettelle (eds.) Berlin: Springer-Verlag, pp. 257–75.

Gwertzman, Bernard. 1985. Reagan said to offer to share "benefits" of "Star Wars" plan. *New York Times*, August 2, p. 4.

——. 1986. Proposals, like weapons, are also proliferating. *New York Times*, April 27, p. E3.

Halloran, Richard. 1986. How leaders think the unthinkable. *New York Times*, September 2, p. A16.

Hardin, Russell. 1983. Unilateral versus mutual disarmament. *Philosophy and Public Affairs* 12, no. 3 (April): 236–54.

Harvard Nuclear Study Group. 1983. *Living with Nuclear Weapons*. New York: Bantam.

Hirshleifer, Jack, and Juan Carlos Martinez Coll. 1987. What strategies can support the evolutionary emergence of cooperation? Center for International and Strategic Affairs Working Paper No. 58. February.

Holloway, David. 1983. *The Soviet Union and the Arms Race*. New Haven, CT: Yale University Press.

Howard, Nigel, 1971. *Paradoxes of Rationality: Theory of Metagames and Political Behavior*. Cambridge, MA: MIT Press.

Ignatius, David. 1985. In Star Wars debate, tactical issues nearly get lost in shuffle. *Wall Street Journal*, October 15, pp. 1, 15.

Iklé, Fred Charles. 1971. *Every War Must End*. New York: Columbia University Press.

——. 1973. Can nuclear deterrence last out the century? *Foreign Affairs* 51, no. 2 (January): 267–85.

Intriligator, Michael D., and Dagobert L. Brito. 1976. Formal models of arms races. *Journal of Peace Science* 2, no. 1 (Spring): 77–88.

—— and ——. 1985. Heuristic decision rules, the dynamics of the arms race, and war initiation. In *Dynamic Models of International Conflict*, Urs Luterbacher and Michael D. Ward (eds.) Boulder, CO: Lynne Rienner, pp. 133–60.

—— and ——. 1986. Mayer's alternative to the I–B model. *Journal of Conflict Resolution* 30, no. 1 (March): 29–31.

Jervis, Robert, Richard Ned Lebow, and Janice Gross Stein (eds.) 1985. *Psychology and Deterrence*. Baltimore: Johns Hopkins University Press.

Kahn, Herman. 1965. *On Escalation: Metaphors and Scenarios*. New York: Praeger.

——. 1984. *Thinking about the Unthinkable in the 1980s*. New York: Simon and Schuster.

Kalai, Ehud. 1981. Preplay negotiations and the Prisoners' Dilemma. *Mathematical Social Sciences* 1: 375–79.

Kavka, Gregory S. 1981. Morality and nuclear politics: lessons of the missile crisis. In *Nuclear Weapons and the Future of Humanity: The Fundamental Questions*, Avner Cohen and Steven Lee (eds.) Totowa, NJ: Rowman and Allanheld, pp. 236–54.

Kerby, William. 1986. The impact of space weapons on strategic stability and the prospects for disarmament: a quantitative analysis. Institut für Friedenforschung und Sicherheitspolitik an der Universität Hamburg. Mimeographed. October.

Kilgour, D. Marc. 1984. Equilibria for far-sighted players. *Theory and Decision* 16, no. 2 (March): 135–57.

———. 1985. Anticipation and stability in two-person non-cooperative games. In *Dynamic Models of International Conflict*, Urs Luterbacher and Michael D. Ward (eds.) Boulder, CO: Lynne Rienner, pp. 26–51.

———, M. De, and Keith W. Hipel. 1986. Conflict analysis using staying power. *Proceedings, IEEE International Conference on Systems, Man, and Cybernetics, Atlanta, GA* (October): 545–59.

———, and Frank C. Zagare. 1987. Holding power in sequential games. *International Interactions* 13, no. 2: 91–114.

Krass, Allan S. 1985. *Verification: How Much Is Enough?* Lexington, MA: Lexington.

Langlois, Jean-Pierre. 1987. Mathematical modeling of deterrence and crisis dynamics. Department of Mathematics, San Francisco State University. Mimeographed.

Lebow, Richard Ned. 1987. *Nuclear Crisis Management: A Dangerous Illusion*. Ithaca, NY: Cornell University Press.

Leininger, Wolfgang. 1987. Escalation and cooperation in international conflicts: the dollar auction revisited. University of Bonn. Mimeographed.

Lipman, Barton L. 1986. Cooperation among egoists in Prisoners' Dilemma and Chicken games. *Public Choice* 51, no. 3: 315–31.

Luce, R. Duncan, and Howard Raiffa. 1957. *Games and Decisions: Introduction and Survey*. New York: Wiley.

Lynn, William J. 1985. Existing U.S.-Soviet confidence-building measures. In *Preventing Nuclear War: A Realistic Approach*, Barry M. Blechman (ed.) Bloomington, IN: Indiana University Press, pp. 24–51.

MccGwire, Michael. 1987. *Military Objectives in Soviet Foreign Policy*. Washington, DC: Brookings.

McGinnis, Michael. 1986. Issue linkage and the evaluation of international cooperation. *Journal of Conflict Resolution* 30, no. 1 (March): 141–70.

Majeski, Stephen J. 1984. Arms races as iterated Prisoners' Dilemmas. *Mathematical Social Sciences* 7, no. 3 (June): 253–66.

Maoz, Zeev. 1983. Resolve, capabilities, and the outcomes of interstate disputes, 1816–1976. *Journal of Conflict Resolution* 27, no. 2 (June): 195–229.

———. 1985. Decision-theoretic and game-theoretic models of International conflict. In *Dynamic Models of International Conflict*, Urs Luterbacher and Michael D. Ward (eds.) Boulder, CO: Lynne Rienner, pp. 76–111.

——— and Dan S. Felsenthal. 1987. Self-binding commitments, the inducement of trust, social choice, and the theory of international cooperation. *International Studies Quarterly* 31, no. 2 (June): 177–200.

Marsh, Gerald E. 1985. SDI: the stability question. *Bulletin of the Atomic Scientists* 41, no. 9 (October): 23–24.

Marshall, Eliot. 1986. If terrorists go nuclear. *Science* 233, no. 4760 (11 July): 148–49.

Martel, William C., and Paul L. Savage. 1986. *Strategic Nuclear War: What the Superpowers Target and Why*. Westport, CT: Greenwood.

Maschler, Michael. 1966. A price leadership method for solving the inspector's non-constant-sum game. *Naval Research Logistics Quarterly* 13, no. 1 (March): 11–33.

———. 1967. The inspector's non-constant-sum game: its dependence on a system of detectors. *Naval Research Logistics Quarterly* 14, no. 3 (September): 275–90.

Mayer, Thomas F. 1986. Arms races and war initiation: some alternatives to the Intriligator-Brito model. *Journal of Conflict Resolution* 30, no. 1 (March): 3–28.

Max, C., 1986. Deployment stability of strategic defenses. McLean, VA: Jason, Mitre Corporation. October.

Mendolovitz, Saul H. 1985. In "Forum: is there a way out?" *Harper's* 270, no. 1621 (June): 35–40.

Miller, Steven E., and Stephen Van Evera (eds.) 1986. *The Star Wars Controversy: An International Security Reader.* Princeton, NJ: Princeton University Press.

Mishal, Shaul, David Schmeidler, and Itai Sened. 1987. Israel and the PLO: a game with differential information. Working Paper No. 8–87, Foerder Institute for Economic Research, University of Tel-Aviv. May.

Mohr, Charles. 1983. U.S. urged to shift A-attack policy. *New York Times*, November 23, p. A7.

———. 1986. Star Wars planners are digging themselves in. *New York Times*, April 20, p. E4.

Molander, Per. 1985. The optimal level of generosity in a selfish, uncertain environment. *Journal of Conflict Resolution* 29, no. 4 (December): 611–18.

Morgan, Patrick M. 1986. New directions in deterrence theory. In *Nuclear Weapons and the Future of Humanity: The Fundamental Questions*, Avner Cohen and Steven Lee (eds.) Totowa, NJ: Rowman and Allanheld, pp. 169–89.

Morrow, James D. 1984. A twist of truth: a re-examination of the effect of arms races on the occurrence of war. Department of Political Science, University of Michigan. Mimeographed.

———. 1986. A spatial model of international conflict. *American Political Science Review* 80, no. 4 (December): 1131–50.

———. 1987. A limited information model of crisis bargaining. Department of Political Science, University of Michigan. Mimeographed.

Nacht, Michael. 1985. *The Age of Vulnerability: Threats to the Nuclear Stalemate.* Washington, DC: Brookings.

Nalebuff, Barry. 1986. Brinkmanship and nuclear deterrence: the neutrality of escalation. *Conflict Management and Peace Science* 9, no. 2 (Spring): 19–30.

Nash, John, 1951. Non-cooperative games. *Annals of Mathematics* 54: 286–95.

New York Times. 1985. Nixon says he considered using atomic weapons on 4 occasions. July 22, p. A12.

Nincic, Miroslav. 1986. Can the U.S. trust the U.S.S.R.? *Scientific American* 254, no. 4 (April): 33–41.

Niou, Emerson M. S., and Peter C. 1986. A theory of the balance of power in international systems. *Journal of Conflict Resolution* 30, no. 4 (December): 685–715.

——— and ———. 1987. Preventive war and the balance of power: a game-theoretic approach. *Journal of Conflict Resolution* 31, no. 3 (September): 387–419.

Nye, Joseph S., Jr. 1987. Nuclear learning and U.S.-Soviet security regimes. *International Organization* 41, no. 3 (Summer): 369–402.

Nye, Joseph S., Jr., Graham T. Allison, and Albert Carnesale. 1985. Analytic conclusions: hawks, doves, and owls. In *Hawks, Doves, and Owls: An Agenda for Avoiding Nuclear War*, Allison, Carnesale, and Nye (eds.) New York: Norton, pp. 205–22.

Office of Technology Assessment. 1985. *Ballistic Missile Defense Technologies: Summary.* Washington, DC: Office of Technology Assessment. September.

O'Flaherty, Brendan. 1985. *Rational Commitment: A Foundation for Macroeconomics*. Durham, NC: Duke University Press.

O'Neill, Barry. 1986. International escalation and the dollar auction. *Journal of Conflict Resolution* 30, no. 1 (March): 33–50.

——. 1987. A measure for crisis instability, with applications to space-based antimissile defense. *Journal of Conflict Resolution* 31, no. 4 (December).

——. 1988. Game theory and the study of the deterrence of war. In *Perspectives on Deterrence*, Robert Axelrod, Robert Jervis, Roy Radner, and Paul Stern (eds.) Washington, DC: National Academy of Sciences.

Parks, Roger B. 1985. What if "Fools Die"? A comment on Axelrod. *American Political Science Review* 79, no. 4 (December): 1173–74.

Payne, Keith B. 1986. *Strategic Defense: "Star Wars" in Perspective*. Boston: Hamilton.

Petersen, Walter J. 1986. Deterrence and compellence: a critical assessment of conventional wisdom. *International Studies Quarterly* 30, no. 3 (September): 269–94.

Pillar, Paul R. 1983. *Negotiating Peace: War Termination as a Bargaining Process*. Princeton, NJ: Princeton University Press.

Potter, William C. 1985. Introduction. In *Verification and Arms Control*, William C. Potter (ed.) Lexington, MA: Lexington, pp. 1–5.

Powell, Robert. 1987. Crisis bargaining, escalation, and MAD. *American Political Science Review* 81, no. 3 (September): 717–35.

Quester, George H. 1986. *The Future of Nuclear Deterrence*. Lexington, MA: Lexington.

Raiffa, Howard. 1982. *The Art and Science of Negotiation*. Cambridge, MA: Harvard University Press.

Rapoport, Anatol. 1966. *Two-Person Game Theory: The Essential Ideas*. Ann Arbor, MI: University of Michigan Press.

——, Melvin J. Guyer, and David G. Gordon. 1976. *The 2×2 Game*. Ann Arbor, MI: University of Michigan Press.

Richelson, Jeffrey. 1985. Technical collection and arms control. In *Verification and Arms Control*, William C. Potter (ed.) Lexington, MA: Lexington, pp. 169–216.

Roth, Alvin E. (ed.) 1985. *Game-Theoretic Models of Bargaining*. Cambridge, UK: Cambridge University Press.

Rowell, William F. 1986. *Arms Control Verification: A Guide to Policy Issues for the 1980s*. Cambridge, MA: Ballinger.

Russett, Bruce. 1983. *The Prisoners of Insecurity: Nuclear Deterrence, the Arms Race, and Arms Control*. New York: Freeman.

Sanger, David E. 1985. Many hesitant to share "Star Wars." *New York Times*, November 30, p. 3.

Saperstein, Alvin M., and Gottfried Mayer-Kress. 1987. A nonlinear dynamical model of the impact of S.D.I. on the arms race. Los Alamos, NM: Center for Nonlinear Studies, Los Alamos National Laboratory.

Schelling, Thomas C. 1960. *The Strategy of Conflict*. Cambridge, MA: Harvard University Press.

——. 1966. *Arms and Influence*. New Haven, CT: Yale University Press.

Scowcroft, Brent, John Deutsch, and R. James Woolsey. 1987. A way out of Reykjavik. *New York Times Magazine*, January 25, pp. 40–42, 78–84.

Scribner, Richard A., Theodore J. Ralston, and William D. Metz. 1985. *The*

Verification Challenge: Problems and Promise of Strategic Nuclear Arms Control Verification. Boston: Birkhaüser.

Selten, Reinhard. 1975. Reexamination of the perfectness concept for equilibrium points in extensive games. *International Journal of Game Theory* 4, no. 1: 25–55.

Sexton, Thomas R., and Dennis R. Young. 1985. Game tree analysis of international crises. *Journal of Policy Analysis and Management* 4, no. 3: 354–69.

Shepherd, William G. 1986. *The Ultimate Deterrent: Foundations of US–USSR Security under Stable Competition.* New York: Praeger.

Shubik, Martin. 1971. The dollar auction game: a paradox in noncooperative behavior and escalation. *Journal of Conflict Resolution* 15, no. 1 (March): 109–11.

———. 1982. *Game Theory in the Social Sciences: Concepts and Solutions.* Cambridge, MA: MIT Press.

Simon, Paul. 1986. Quoted in "Nuclear freeze debate spawns metaphor war." *New York Times*, April 25, p. A14.

Siverson, Randolph M. 1986. The conflict spiral, arms races and Soviet-American relations. Department of Political Science, University of California, Davis. Mimeographed.

Smith, R. Jeffrey. 1985. Allegations of cheating endanger arms talks. *Science* 227 (8 March): 1180–81.

Snidal, Duncan. 1985. The game *theory* of international politics. *World Politics* 28, no. 1 (October): 25–57.

Snow, Donald M. 1983. *The Nuclear Future: Toward a Strategy of Uncertainty.* University, AL: University of Alabama Press.

———. 1987. *The Necessary Peace: Nuclear Weapons and Superpower Relations.* Lexington, MA: Lexington.

Snyder, Glenn H., and Paul Diesing. 1977. *Conflict among Nations: Bargaining, Decision Making, and System Structure in International Crises.* Princeton, NJ: Princeton University Press.

Sorensen, Theodore C. 1965. *Kennedy.* New York: Harper and Row.

Stefanski, Jacek. 1987. Deterrence and the negotiated resolution of a crisis. Systems Research Institute, Polish Academy of Sciences, Warsaw. Mimeographed.

Stein, Janice Gross. 1975. War termination and conflict resolution, or how wars should end. *Jerusalem Journal of International Relations* 1, no. 1 (Fall): 1–27.

Stever, H. Guyford, and Heinz R. Pagels (eds.) 1986. *The High Technologies and Reducing the Risk of War, Annals of the New York Academy of Sciences*, vol. 489. New York: New York Academy of Sciences.

Taylor, Michael. 1976. *Anarchy and Cooperation.* London: Wiley Ltd.

Thomas, L. C. 1987. Using game theory and its extensions to model conflict. In *Analyzing Conflict and Its Resolution: Some Mathematical Contributions*, P.G. Bennett (ed.) Oxford, UK: Clarendon, pp. 3–22.

Thompson, E. P. (ed.) 1985. *Star Wars.* New York: Pantheon.

Trachtenberg, Marc. 1985. The influence of nuclear weapons in the Cuban missile crisis. *International Security* 10, no. 1 (Summer): 137–63.

Tsipis, Kosta. 1987. Arms control pacts can be verified. *Discover* 8, no. 2 (April): 79–83.

———, David Hafemeister, and Penny Janeway (eds.) 1986. *Arms Control Verification: The Technologies That Make It Possible.* Washington, DC: Pergamon Brassey's.

Union of American Hebrew Congregations. 1981. *The Torah: A Modern Commentary*. New York: Union of American Hebrew Congregations.

Ury, William. 1985. *Beyond the Hotline: How Crisis Control Can Prevent Nuclear War*. Boston: Houghton Mifflin.

Vick, Alan J., and James A. Thompson. 1985. The military significance of restrictions on the operations of nuclear forces. In *Preventing Nuclear War: A Realistic Approach*, Barry M. Blechman (ed.) Bloomington, IN: Indiana University Press, p. 123.

von Neumann, John, and Oskar Morgenstern. 1953. *Theory of Games and Economic Behavior*, 3rd ed. Princeton, NJ: Princeton University Press.

von Stackelberg, H. 1934. *Marketform und Gleichgewicht*. Berlin: Julius Springer.

Wagner, R. Harrison. 1982. Deterrence and bargaining. *Journal of Conflict Resolution* 36, no. 2 (June): 329–58.

——. 1983. The theory of games and the problem of international cooperation. *American Political Science Review* 77, no. 2 (June): 330–46.

——. 1986. The theory of games and the balance of power. *World Politics* 38, no. 4 (July): 546–76.

Wang, Muhong, Keith W. Hipel, and Niall M. Fraser. 1986. Hypergame analysis of the Falkland islands crisis. Department of Systems Design Engineering, University of Waterloo. Mimeographed. October.

Weapons in Space (1985), Vols. I and II. In *Daedalus* 114, nos. 2 and 3 (Spring and Summer).

Wells, Samuel F., Jr., and Robert S. Litwak (eds.). 1987. *Strategic Defenses and Soviet–American Relations*. Cambridge, MA: Ballinger.

Wieseltier, Leon. 1985. When deterrence fails. *Foreign Affairs* 63, no. 4 (Spring): 827–47.

Wilkening, Dean, and Kenneth Watman. 1986. Strategic defenses and first-strike stability. Santa Monica, CA: RAND Corporation. November.

Witt, Ulrich. 1986. Evolution and stability of cooperation without enforceable contracts. *Kyklos* 39, fasc. 2: 245–66.

Wittman, Donald. 1979. How a war ends: a rational model approach. *Journal of Conflict Resolution* 23, no. 4 (December): 743–63.

——. 1985. Arms control, verification and other games involving signaling and detection. Working Paper No. 35, Seminar in Applied Economics/Public Finance, University of California, Santa Cruz. March.

Zagare, Frank C. 1984. Limited-move equilibria in 2×2 games. *Theory and Decision* 16, no. 1 (January): 1–19.

——. 1985. Toward a reformulation of the theory of mutual deterrence. *International Studies Quarterly* 29, no. 2 (June): 155–69.

——. 1987. *The Dynamics of Deterrence*. Chicago: University of Chicago Press.

Zraket, Charles A. 1984a. Strategic command, control, communications, and intelligence. *Science* 224, no. 4655 (22 June): 1306–11.

——. 1984b. Response to Brams's letter, *Science* 226, no. 4676 (16 November): 782.

Index

DATE DUE

NOV 0 6 1994			

Demco, Inc. 38-293